HOW TO MAKE COMPUTER-CONTROLLED ROBOTS

Tony Potter

CONTENTS

Robot designed by Tony Potter and Chris Oxlade
Robot program by Chris Oxlade

Illustrated by Jeremy Gower

Additional Illustrations by Simon Roulstone, Chris Lyon, Jeremy Banks, Graham Round, Diane Potter, Hussein Hussein.

Technical consultants: Colin Motteram, John Hawkins and Bill Pinder.

How to use this book

In this book you can find out about computer-controlled robots and how to make them. There are step-by-step instructions showing how to build a robot which moves around on wheels and picks things up with its gripper. This is shown below, painted orange.

Building the robot

The book is divided into sections, showing how to make one part of the robot at a time. After each section there are tests and checks to make sure the part works properly before you go on. At the back of the book there are templates to copy and use to make all the robot parts.

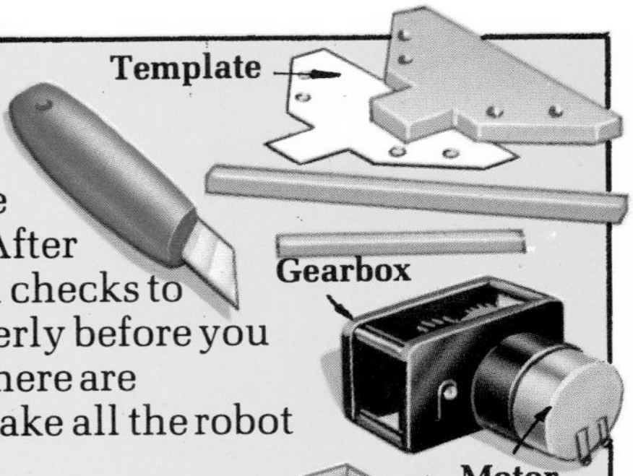

The robot is designed so that you can make different versions. If you like, you can make a robot vehicle like the blue one below (this is the simplest version). Miss out pages 12-19 to do this. Alternatively, you could make a stationary arm robot by leaving off the wheels and wheel motors. There are also customizing ideas on pages 20-21, so you can make different robots from the same design.

All versions of the robot work with the computers shown above. You need to buy an extra part, called an interface*, for the Spectrum. There is a program at the back of the book for these computers, and a special set of robot instructions, called Robotrol, which you can use to control the robot. There are also test programs to check the robot works.

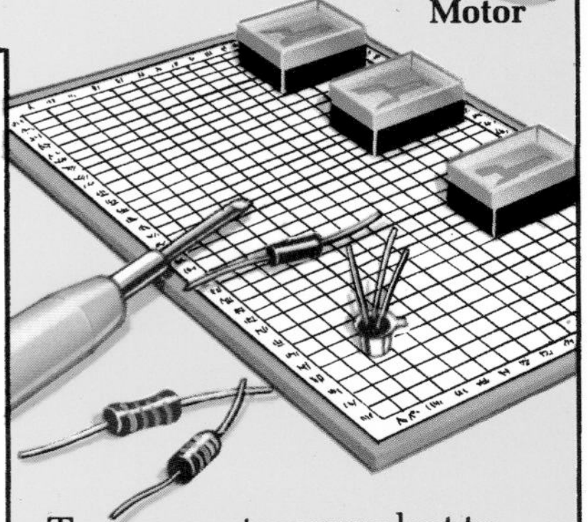

To connect your robot to a computer, there is an electronic circuit to make. This controls the four small battery-powered motors, which drive the robot. You could invent a robot of your own, with up to four motors, and use this circuit to control it.

All robots are precise and accurate machines with lots of moving parts that have to fit together for the robot to work. You need to take your time and follow the instructions in this book very carefully to be successful.

Throughout the book there are lots of practical hints and tips on robot building, and explanations of soldering and electronics. There are also added extras to make for all versions.

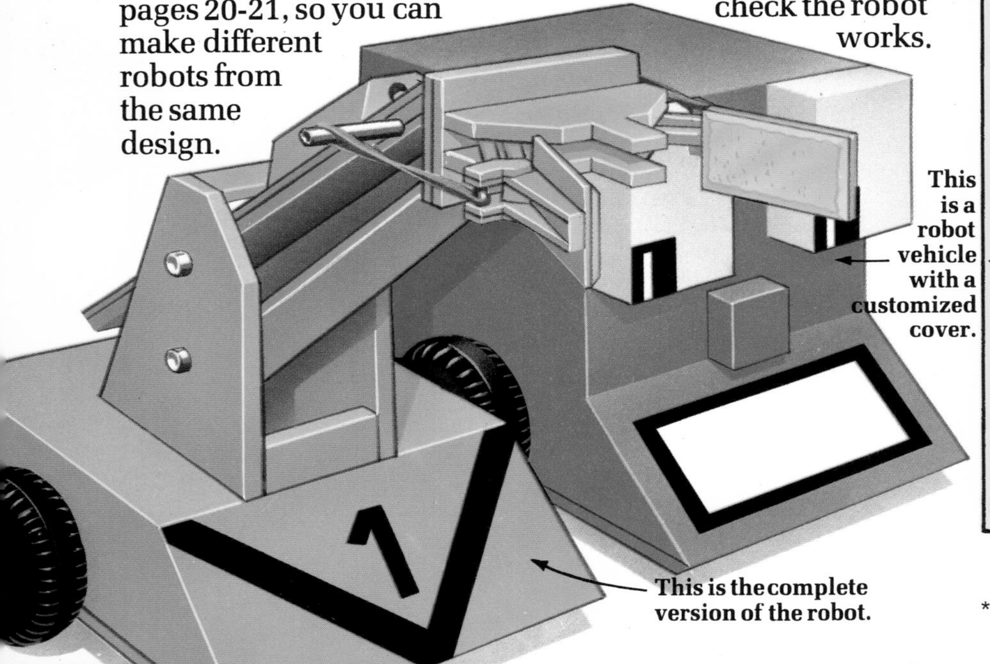

*See pages 9 and 41.

About robots

Real robots are quite different from those of science fiction. They are computer-controlled machines, programmed to use tools or move goods. The study of them is called robotics. Robots are used in industry, sometimes replacing people, but often doing work which is too dangerous for men and women. Others, like the robot in this book and some shown on this page, are for fun or learning about robotics and computers.

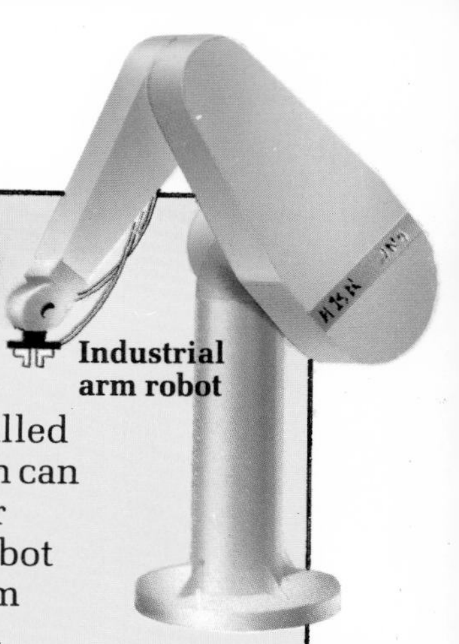

There are two main types of robot. Those with wheels or which move on tracks are called mobile robots. Robots which can hold things are called arm or manipulative robots. The robot to make in this book is an arm and mobile robot combined.

Micro-robots

A micro-robot is a small robot controlled by a home computer. These pictures show some of those available.

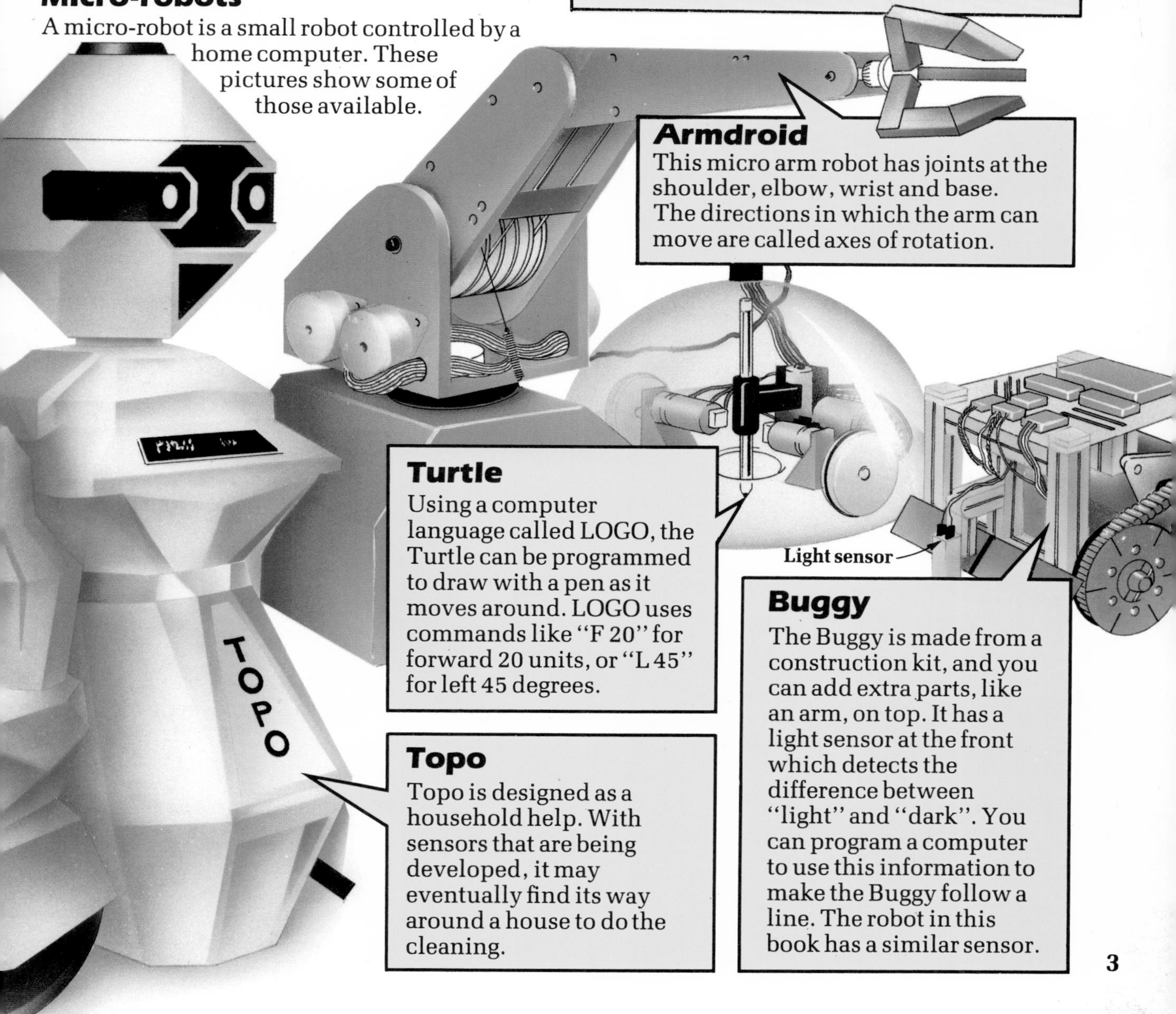

Armdroid

This micro arm robot has joints at the shoulder, elbow, wrist and base. The directions in which the arm can move are called axes of rotation.

Turtle

Using a computer language called LOGO, the Turtle can be programmed to draw with a pen as it moves around. LOGO uses commands like "F 20" for forward 20 units, or "L 45" for left 45 degrees.

Topo

Topo is designed as a household help. With sensors that are being developed, it may eventually find its way around a house to do the cleaning.

Buggy

The Buggy is made from a construction kit, and you can add extra parts, like an arm, on top. It has a light sensor at the front which detects the difference between "light" and "dark". You can program a computer to use this information to make the Buggy follow a line. The robot in this book has a similar sensor.

Building a robot

This cut-away picture shows the finished robot. The instructions in the book show how to make one part at a time, and these are painted in different colours to make it easier to see what they are. You need not build all these parts if you want to make either a robot vehicle or a stationary arm robot.

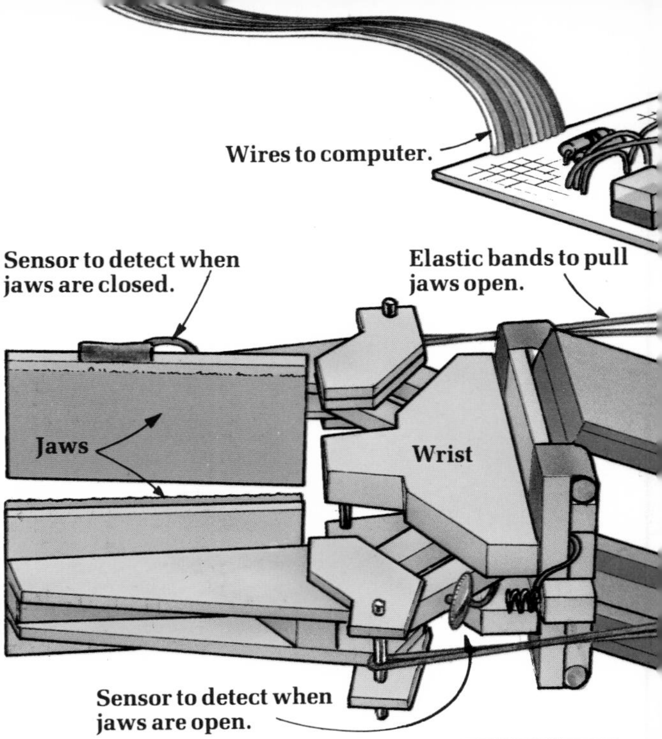

Gripper

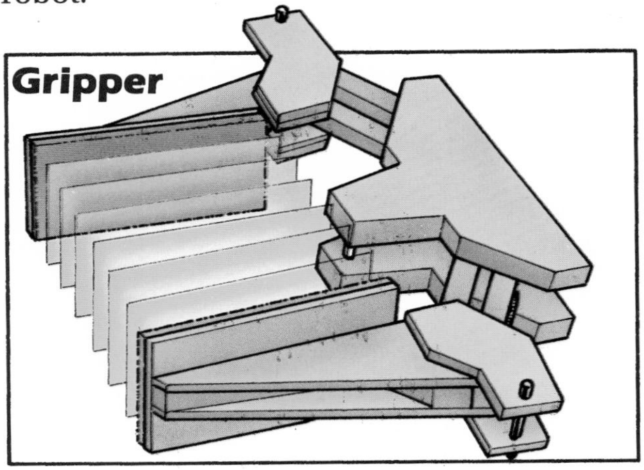

A robot's "hand" is called a gripper. This has two "fingers" or jaws which open and close to pick things up and put them down. The jaws open to about 70mm and can lift something the weight of a small apple.

Sensors

The robot has simple sensors which tell the computer when the arm is fully up or down, whether the jaws are holding something and if the gripper is open or closed. There is also a light sensor you can make to give the robot extremely simple "vision".

Arm

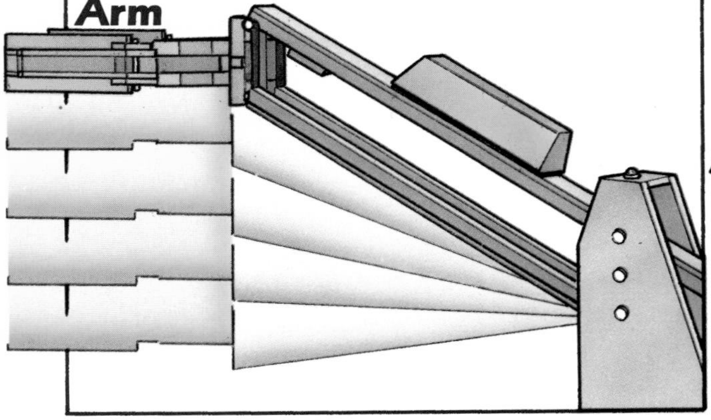

The robot has an arm which moves up and down. It is designed so that as the arm moves, the gripper always stays parallel to the ground. The arm is able to travel about 200mm up and down.

Base

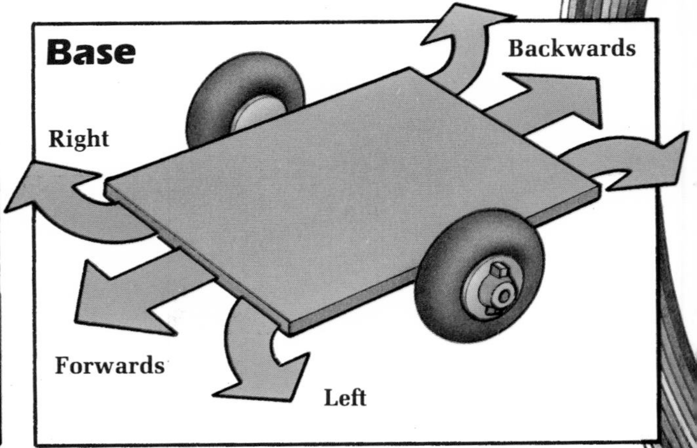

The base of the robot is a flat board with two wheels driven by electric motors. By programming a computer to switch the motors on and off, the robot can be made to go forwards or backwards, left and right.

Electronic switching circuit

You need to build an electronic circuit, called a switching circuit, to control the robot with your computer. This is made of electronic components which switch the robot's motors on and off when signals are sent by the computer.

Shoulder

The shoulder supports the robot's arm. Fishing line control cables, to move the gripper and arm, pass through the shoulder to motors on the base.

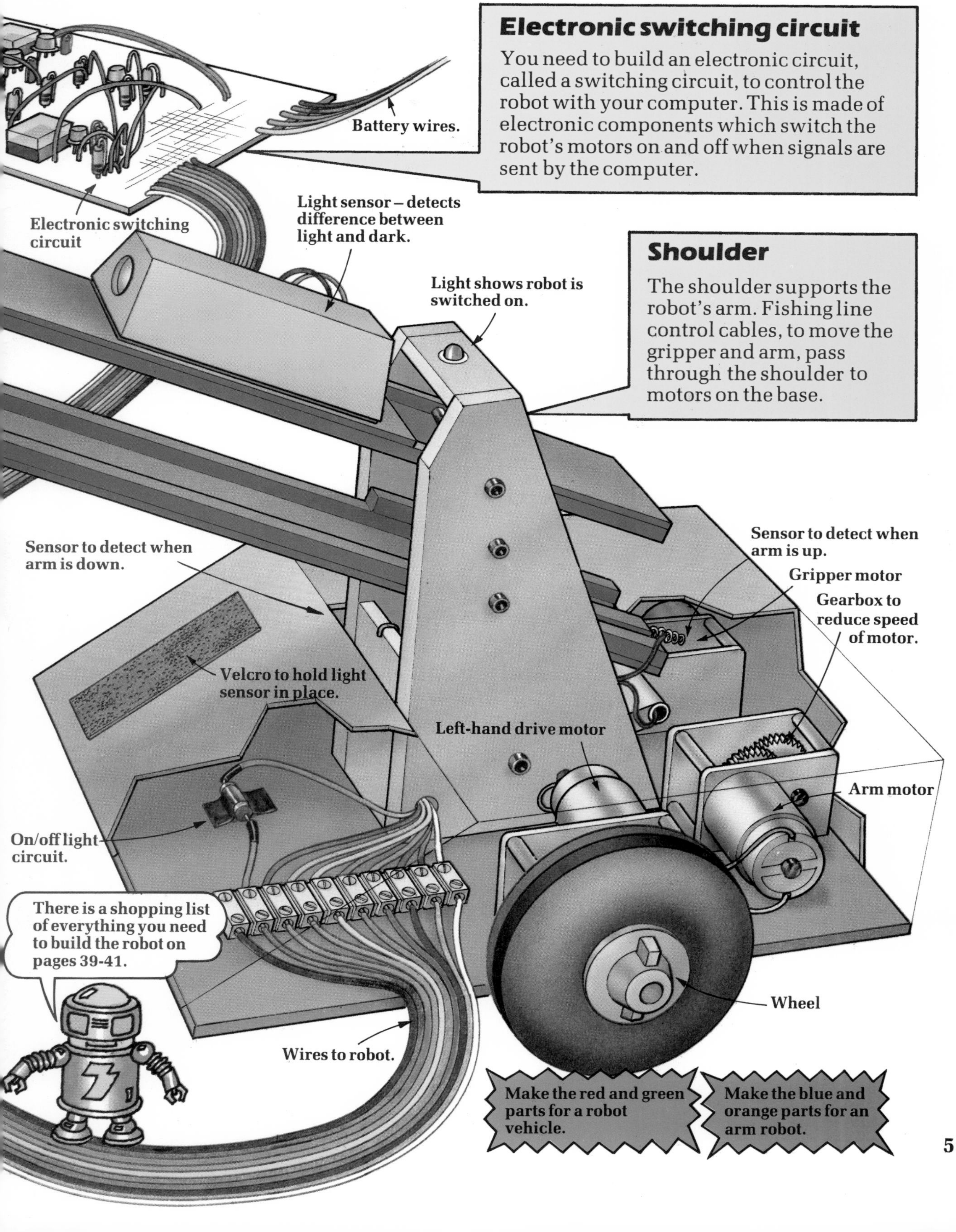

Things you need

Here you can see all the things needed to make and control the robot. Model and hardware shops sell most things, but you need to get the electronic parts from a components shop. Ask in your local TV repair shop to find out where the nearest components supplier is. You can also buy components by post. There is a complete shopping list on pages 39-41, showing sizes and amounts of things to buy, with a list of useful mail-order suppliers.

Tools

The tools you need are shown on the right.

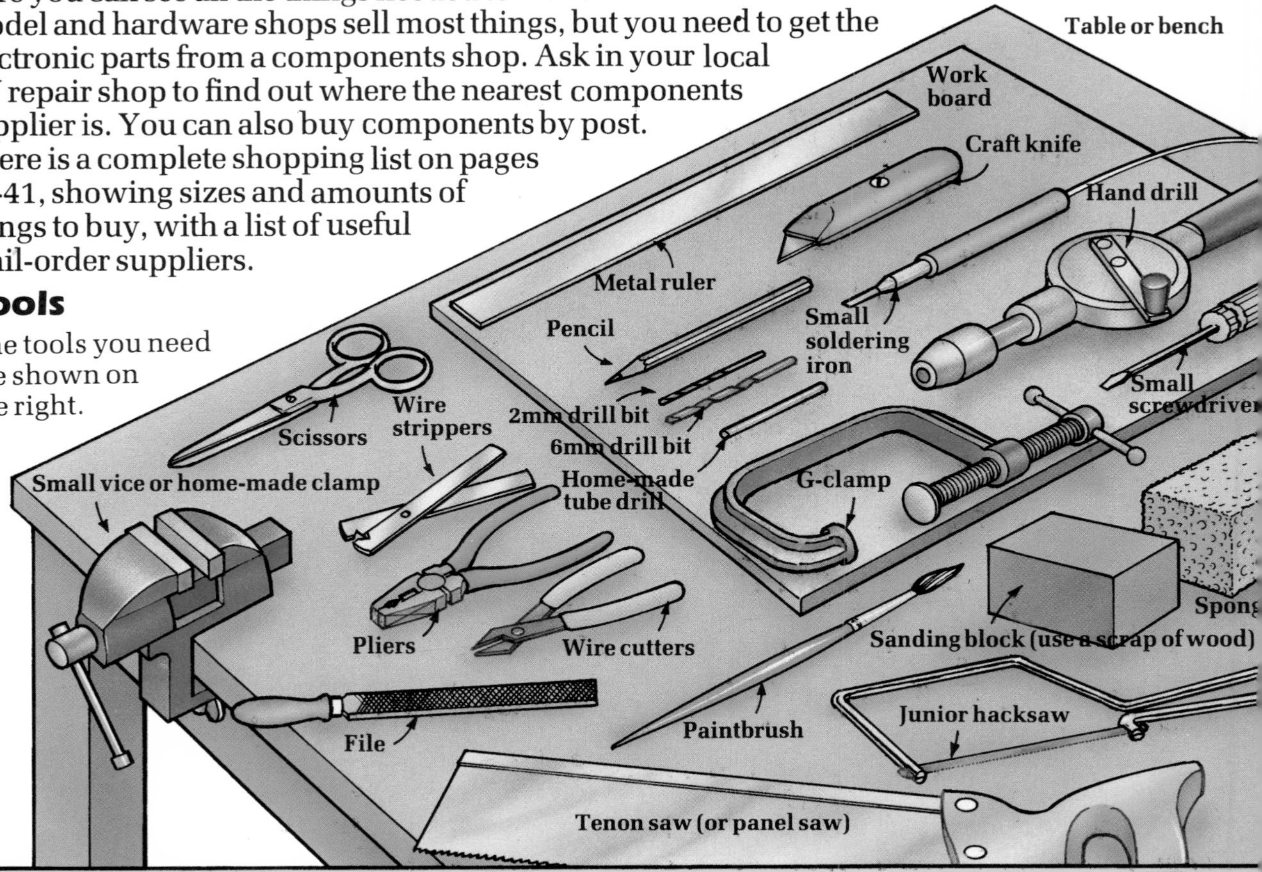

Materials you need

These are the materials needed to construct the robot. It is a good idea to get everything before you begin.

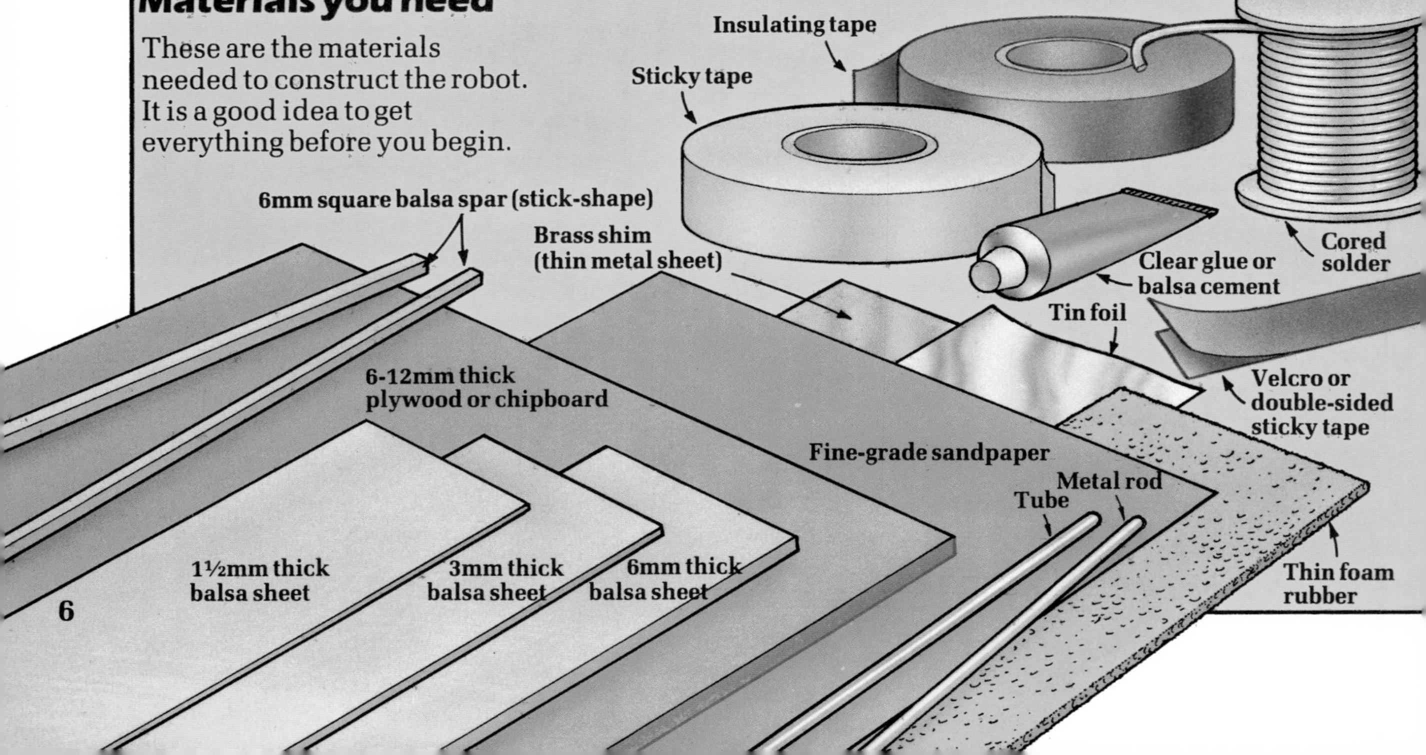

Electronic components

An electronic component is a small device which controls the flow of electricity. There are many kinds, all controlling electricity in different ways. Components are connected together to make a circuit. Those shown here are used to make the switching circuit for the robot. You can find out more about them on pages 22 and 24, and in the shopping list.

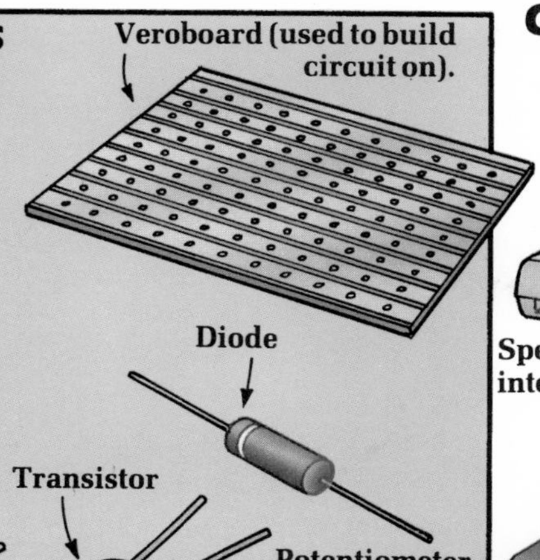

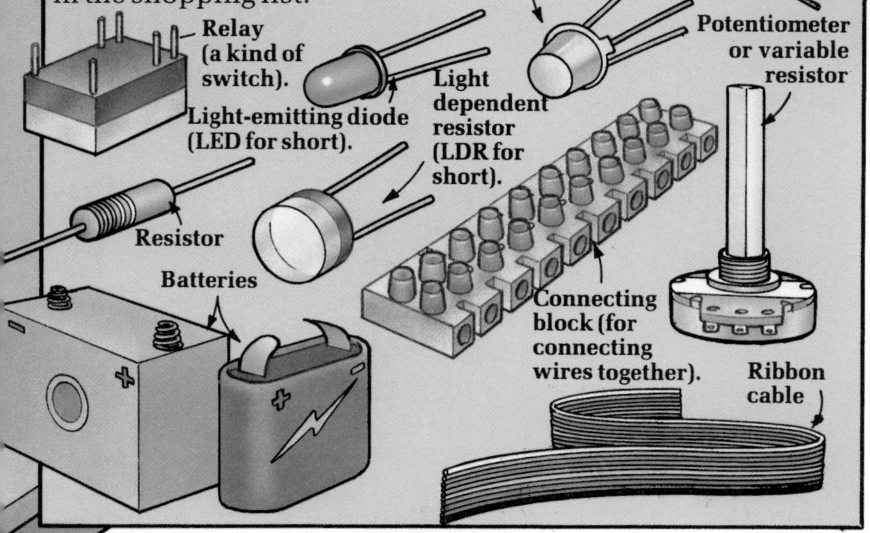

Computer

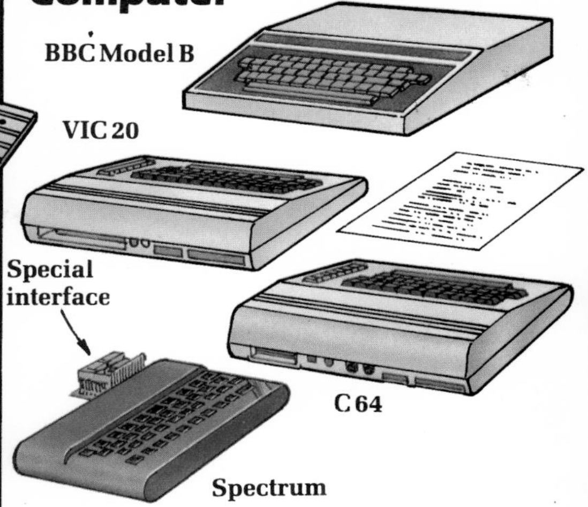

To control a robot you need a home computer with a special socket, or port, called a parallel input/output port. Computers with this port include the Commodore 64, VIC 20 and BBC B. You can buy an extra part for the Spectrum called an I/O interface, which enables it to control a robot. The project robot is designed to work with these computers, using the program on page 36.

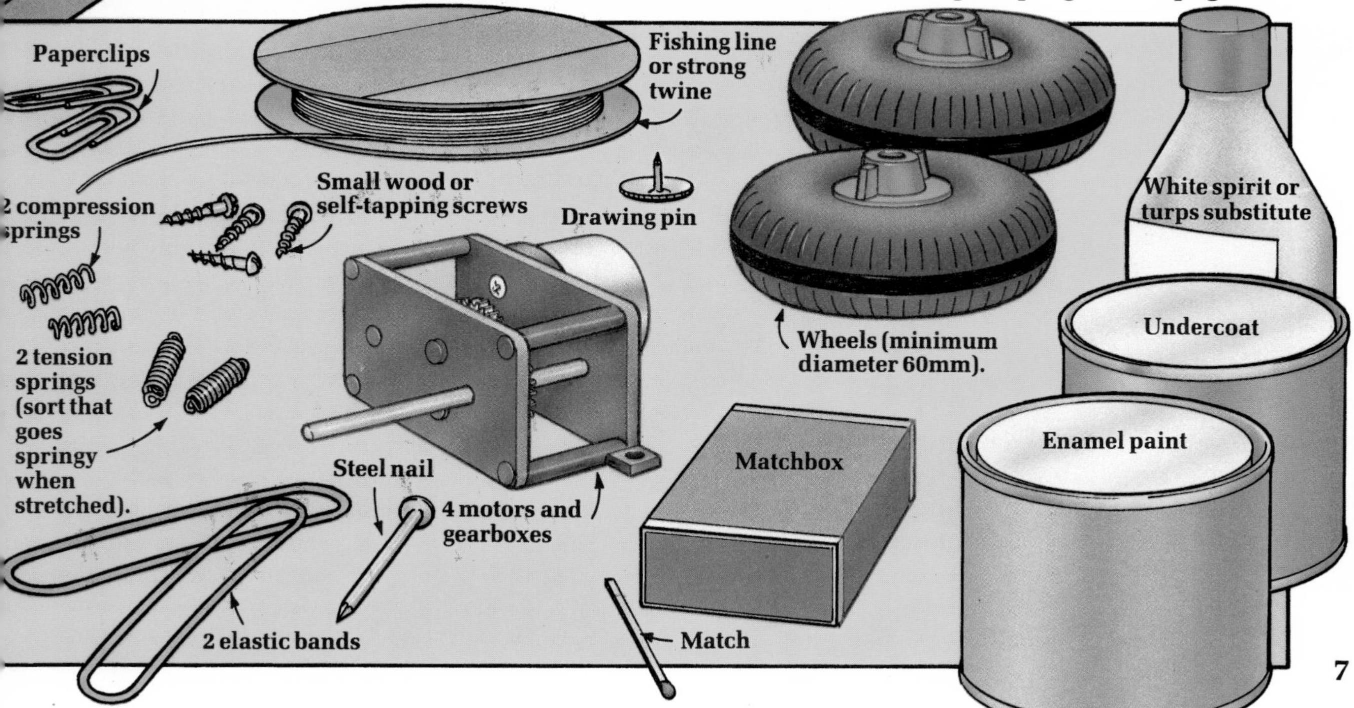

Robot construction tips

To be successful you need to cut and drill all the robot parts very accurately and carefully. This page gives some construction hints and tips. You can also find out how to use the templates at the back of the book.

Home-made clamp

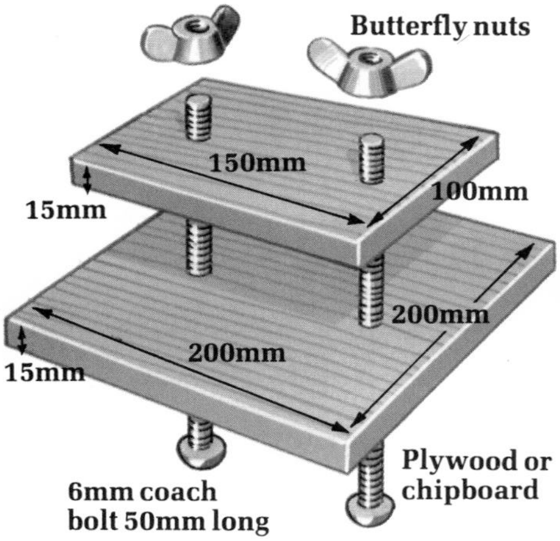

Make a clamp like this to keep materials steady while drilling and cutting. Glue or screw the clamp to your workboard.

Cutting tube

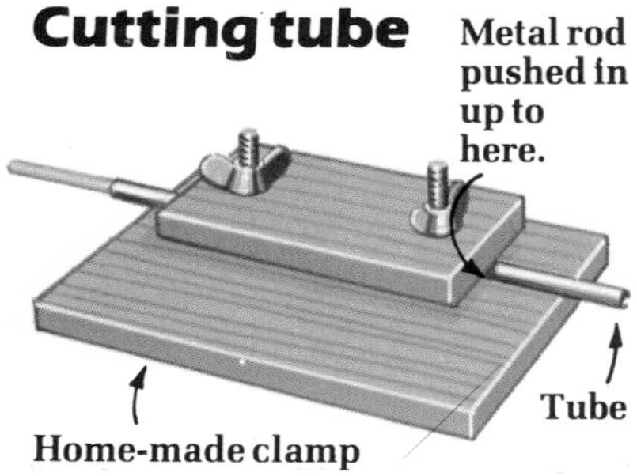

Some parts of the robot are made from metal rod which fits snugly inside thin tube. Clamp the tube or rod in a vice or home-made clamp and cut it with a hacksaw. Push a piece of metal rod inside the clamped part of the tube to avoid crushing it.

Cutting balsa

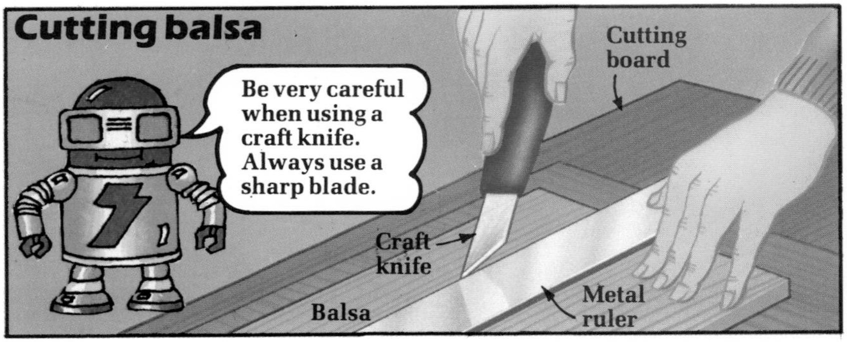

Cut the balsa with a sharp craft knife, using a metal ruler as a guide. Use a piece of hardboard or plywood to work on. Always hold the knife firmly. Stand to one side and cut towards yourself but away from your body.

Tube drill

You can make a home-made bit to drill accurate holes in balsa for tube to go through.

1 Cut off a 100mm length of metal tube.

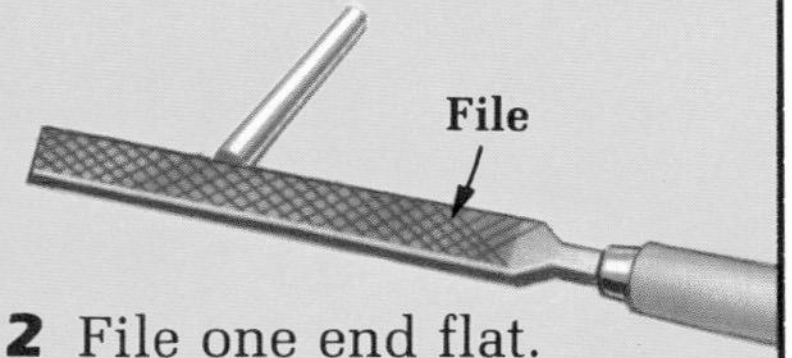

2 File one end flat.

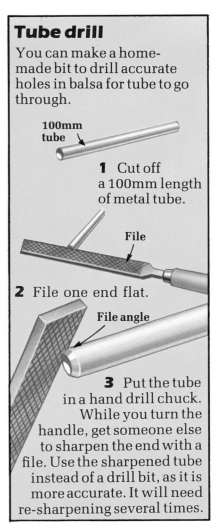

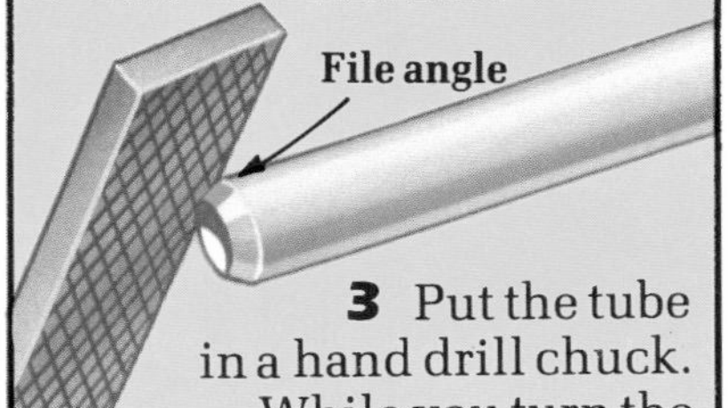

3 Put the tube in a hand drill chuck. While you turn the handle, get someone else to sharpen the end with a file. Use the sharpened tube instead of a drill bit, as it is more accurate. It will need re-sharpening several times.

Test bench reports

- **Motors** ☐
- **Shoulder** ☐
- **Arm** ☐
- **Gripper** ☐
- **Control lines** ☐
- **Cover** ☐
- **Solder check** ☐
- **Circuit** ☐
- **Sensor test** ☐

After building each part of the robot, there are tests and checks to do to make sure it works. You could make a "Test bench report" like this and tick the tests off as you go. If your robot fails any test, there are checks and adjustments to make. You must make sure each part of the robot works properly, or the completed robot will not work either.

Templates

Templates are like patterns, and are used to make all the parts for the robot. Those on pages 42-47 are for all the parts made from sheets of balsa wood. The one on page 41 is for the electronic circuit.

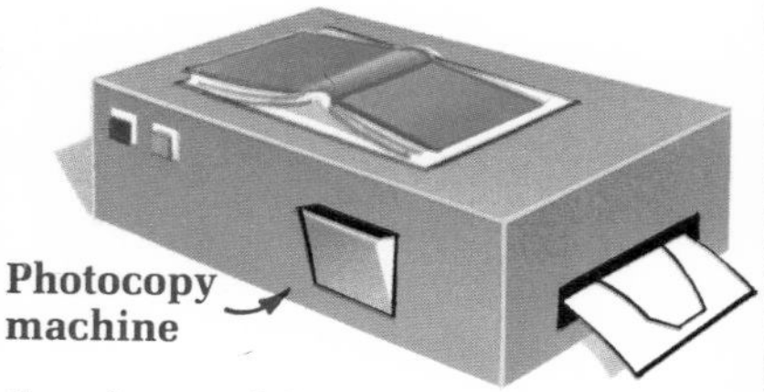

1 If possible, photocopy the template pages of this book. You can find a photocopier in copy shops, some post offices, libraries and railway stations. If you cannot find a copier, carefully trace the templates onto tracing paper. Only do this as a last resort as it is difficult to do accurately.

2 Cut round the templates with scissors, cutting a few millimetres outside the line. Cut each template as you need it, or it might get lost.

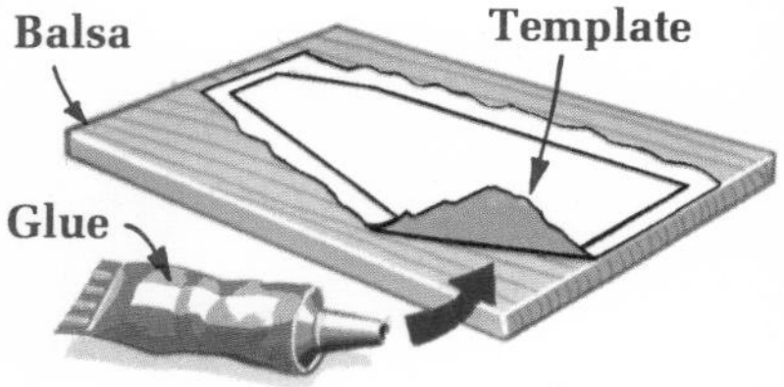

3 Glue each template to the correct thickness of balsa using clear glue or balsa cement. The thickness is printed on each template.

World of robots

Robots are generally deaf, dumb, blind, have no sense of touch, smell or taste, and have no "intelligence" of their own. The computer acts as the robot's "brain", but the robot needs electronic senses, called sensors, for the computer to "know" what the robot is doing. These pictures show robots in use today.

Robot vision

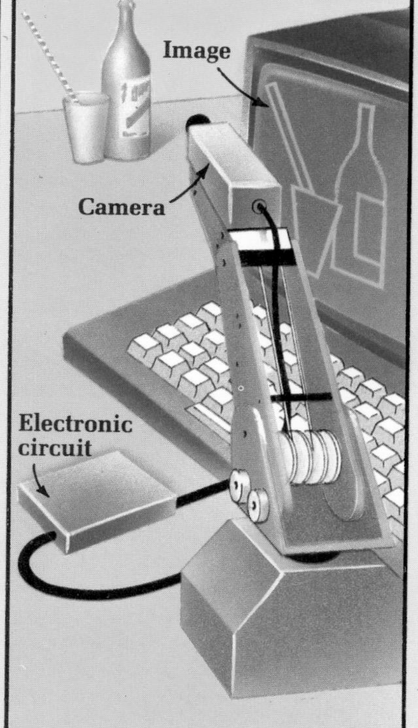

Some robots "see" with a special kind of video camera. The robot's computer is programmed to analyse images from the camera. Images are sent via an electronic circuit which translates them into electrical messages the computer understands.

Factory robots

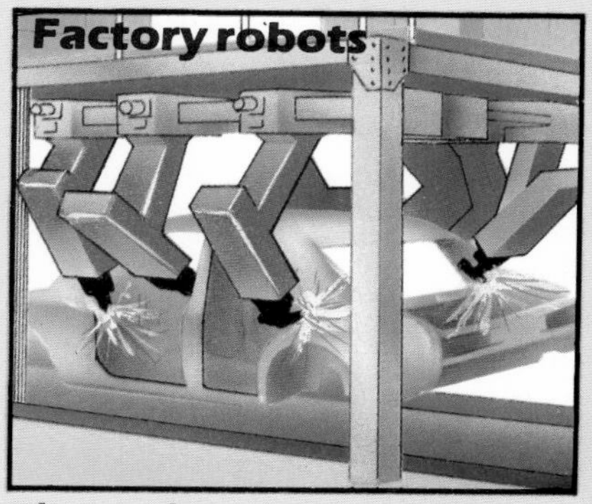

These robots work together to weld a car body as it passes on a conveyor belt. Other robots paint the car later on.

Can robots walk?

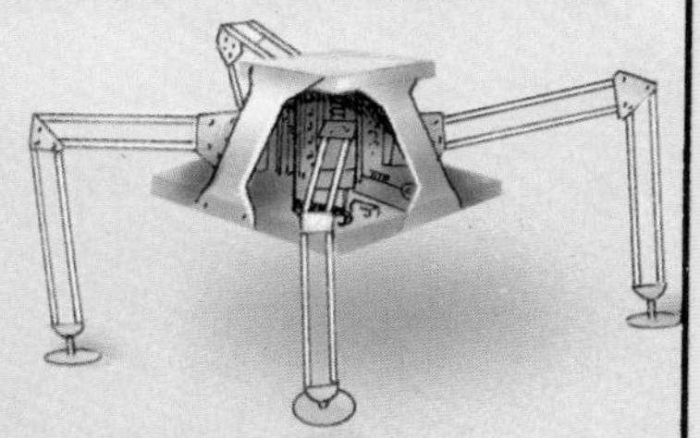

Robots need four or more legs so they always have at least three legs on the ground to balance. Few walking robots are made because it is simpler to use wheels or tracks.

Talking robots

Some robots, like Hero 1, have a speech synthesizer which is programmed to say a limited number of words.

Making the robot base

These instructions show you how to make a mobile robot, either as a base for the project or as a vehicle on its own. Read the instructions and the information on the templates before you begin.

Materials you need

Base template, 6-12mm thick plywood or chipboard, 6mm balsa spar, clear glue, 4 small motors and gearboxes, 8 small screws, battery for motors (e.g. 3V motor needs 3V battery), bell wire or ribbon cable.

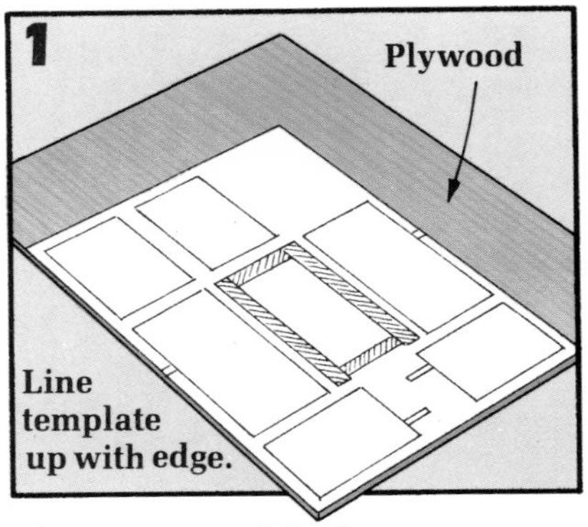

Glue a copy of the base template (page 42) to a wooden board 6-12mm thick, using plenty of clear glue.

Use a tenon or panel saw to cut round the template. Sand the edges with sandpaper wrapped round a block of wood.

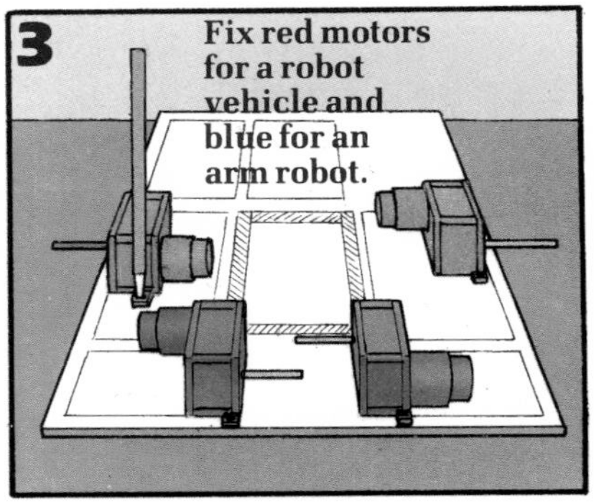

Position the motors and gearboxes as shown on the template, then make a pencil mark through the fixing holes.

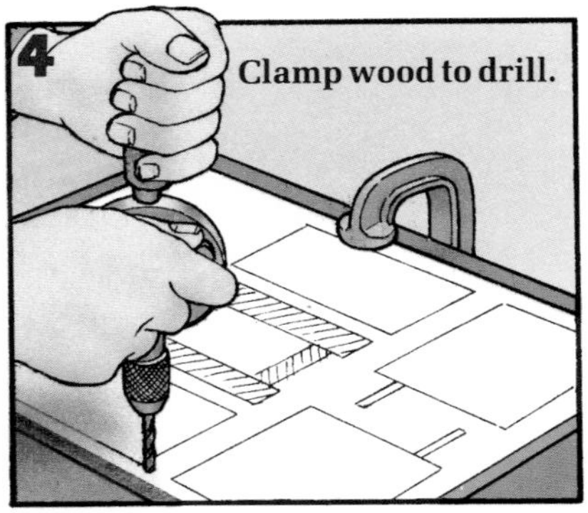

Drill through your pencil marks using a bit slightly smaller than the shafts of the screws for fixing the motors.

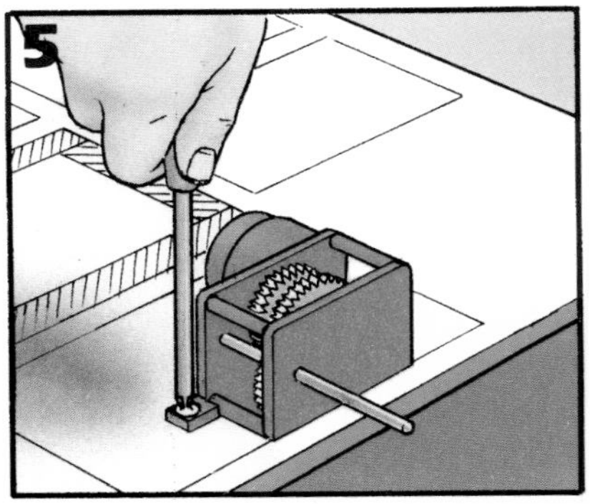

Screw each motor to the base with self-tapping or wood screws. Make sure they are firmly screwed down.

Motors and gearboxes

The robot's wheels, arm and gripper are each driven by a small battery-powered motor. These have a shaft at one end which revolves very quickly – often over 2,000 times per minute. This is too fast to drive the robot, so a gearbox reduces the speed. A series of gears inside the gearbox lock, or mesh, with each other. A small gear spinning fast makes a bigger gear turn more slowly when they mesh. The first gear in the series is turned by a small gear on the motor shaft, and the last turns a shaft coming out of the gearbox. This drives the robot.

This picture shows the insides of the type of motor and gearbox used to illustrate the project. They are sold as kits and are the cheapest and most readily available from model shops, but are difficult to assemble.

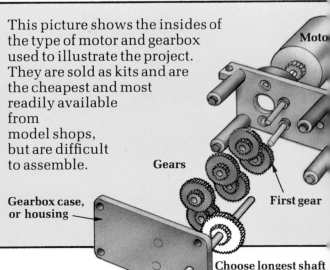

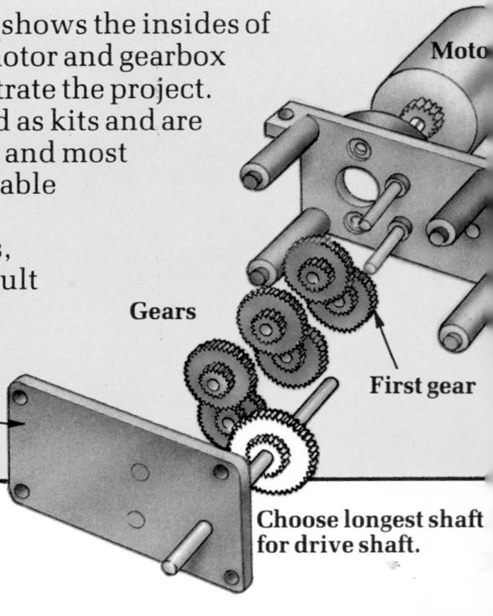

Don't worry if your motors and gears look different from these.

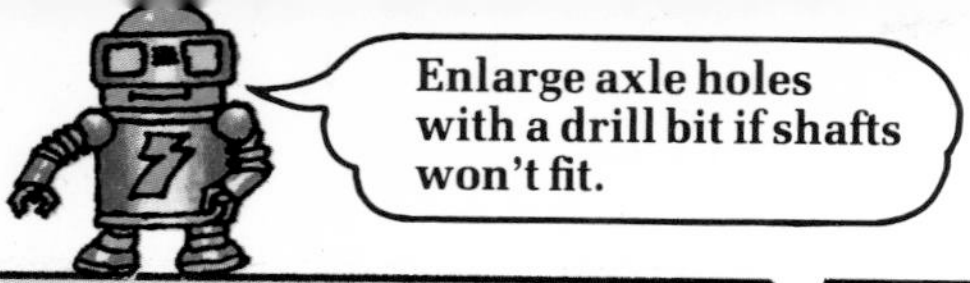

6

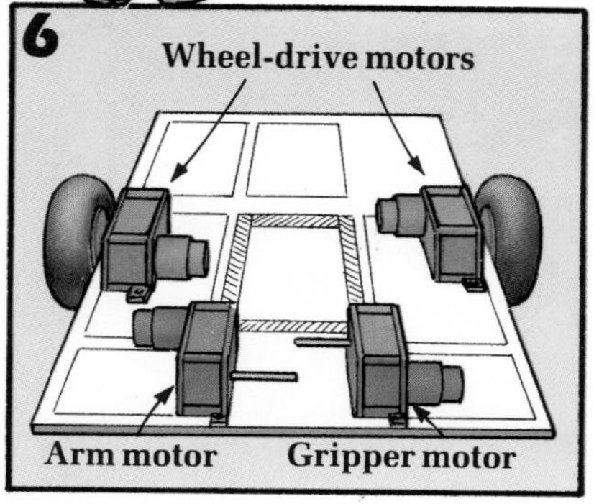

Push the wheels onto the gearbox shafts as shown. If the wheels are loose, wrap tape round the shafts first.

7

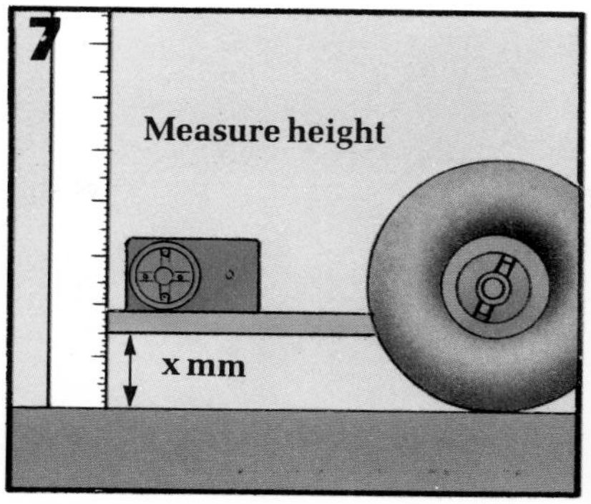

Balance the base to make it level. Measure the distance between ground and base as shown here and write it down.

Sawing tips

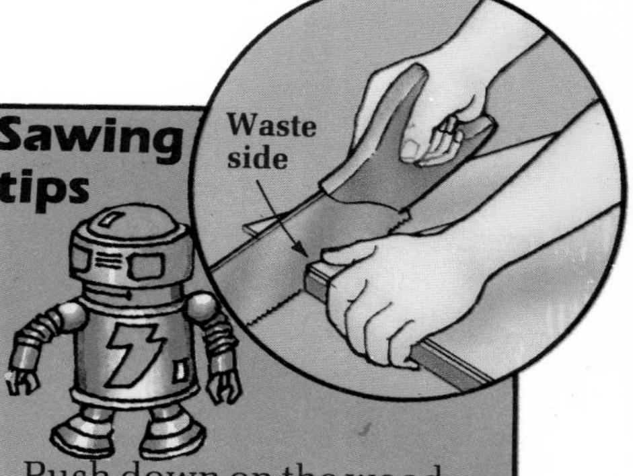

Push down on the wood to steady it. Position the saw on the waste side of the line. Start by sawing downwards, using your thumbnail as a guide. Keep the saw straight.

8

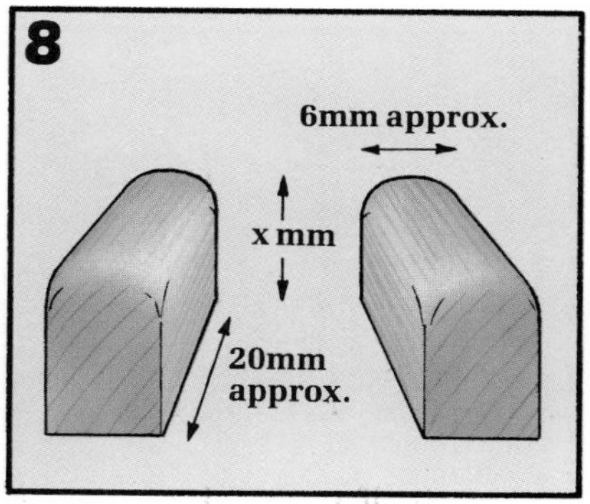

Cut two balsa spars as shown. Sand them down 1mm shorter than the height found in step 7, and round the corners.

9

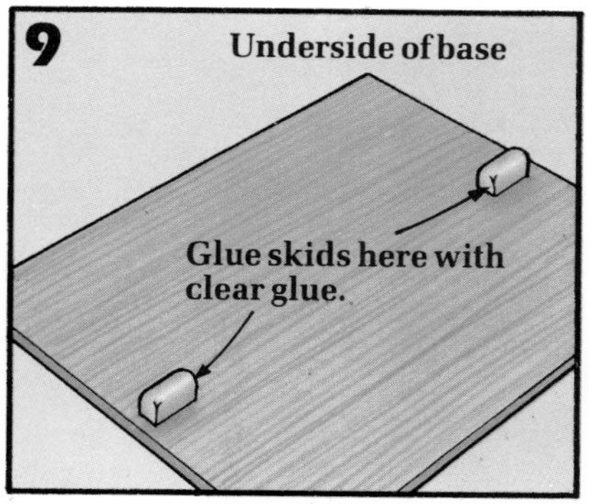

Glue the balsa under the base to act as skids to prevent the robot from tipping up. They work best on a smooth surface.

10

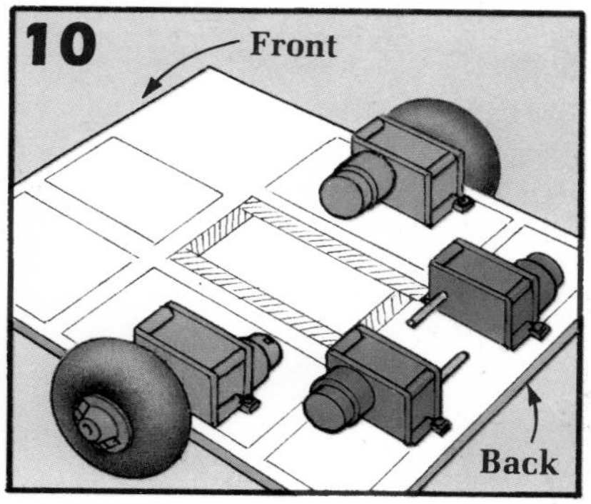

This is how your completed base should look. Sand the skids down if the wheels do not touch the ground.

You could use a motor and ▶ gearbox like the one on the right from a Fischertechnik construction kit.

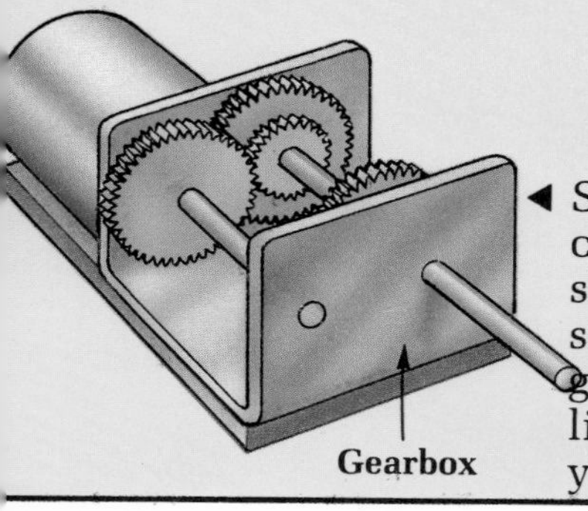

◀ Some electronic component shops have second-hand geared motors like this which you could use.

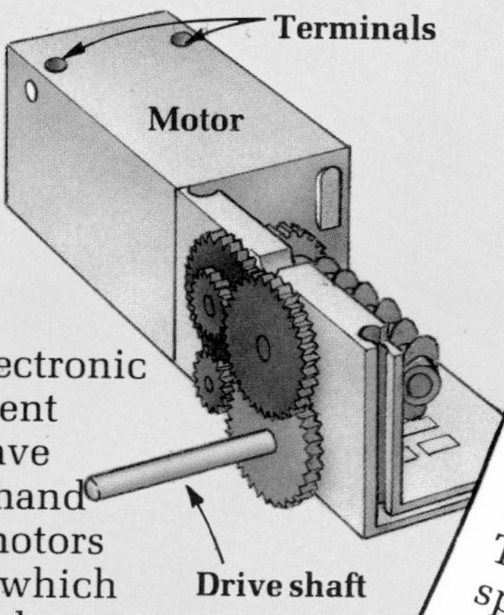

Test bench report

Robot base

Check to see if all your motors work by connecting their

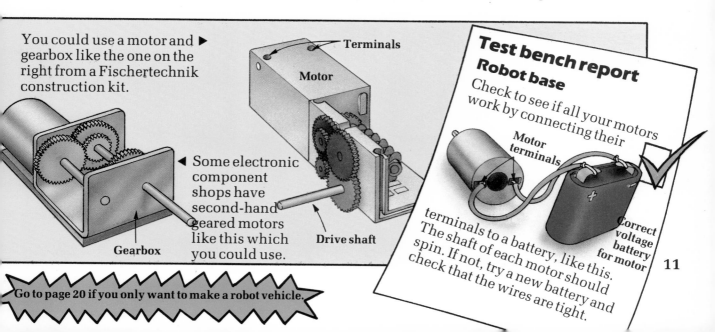

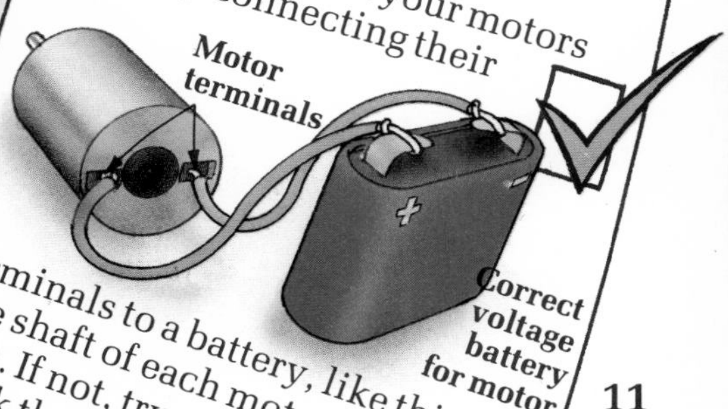

terminals to a battery, like this. The shaft of each motor should spin. If not, try a new battery and check that the wires are tight.

Go to page 20 if you only want to make a robot vehicle.

How to make the shoulder

These pages show how to make a shoulder for the arm robot There are two joints in the shoulder to pivot the arm, letting it move up and down while keeping the gripper parallel with the ground. The shoulder is quite easy to make, but be careful to drill the holes for the pivots accurately.

Materials you need

Shoulder templates A, B, C, D, 90×6mm balsa sheet, 6mm balsa spar, metal or plastic tube (max 6mm diameter), metal rod (to fit snugly inside tube), clear glue or balsa cement. Tool kit including 6mm drill and drill bit or home-made tube drill (same diameter as tube).

Robot joints

Arm robots often have three main parts joined together at pivots. The point where the parts are fixed is called a joint, or axis. These pictures show the special names for the joints and the directions in which they allow the robot to move. Each axis is said to give a robot one degree of freedom because it allows movement in one direction. The project robot has one degree of freedom in its shoulder.

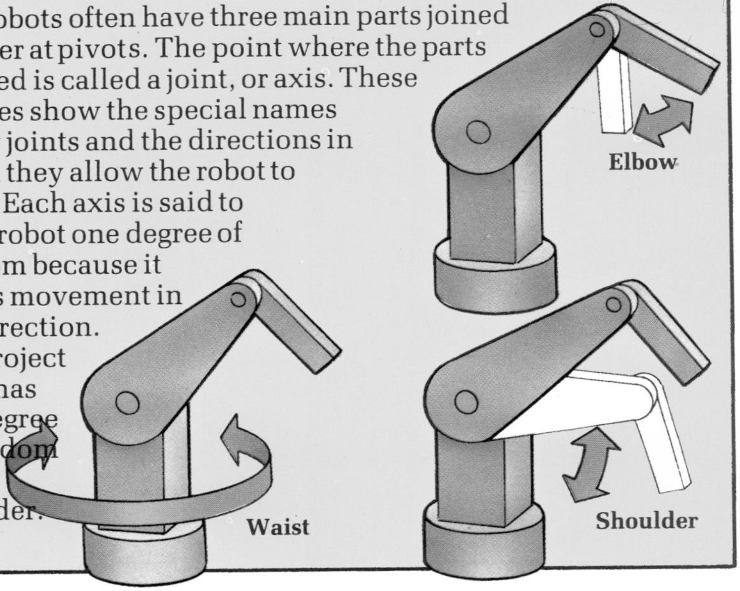

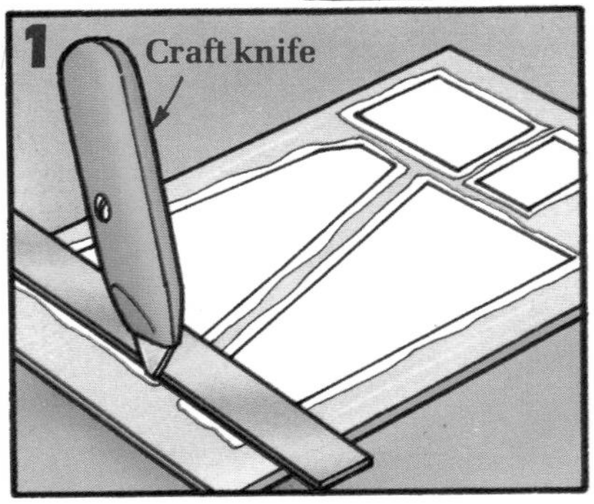

Glue the shoulder templates to 6mm balsa sheet and cut round them with a craft knife. Lightly sand the edges.

Making joints

The robot's moving parts are jointed, or pivoted, with shafts made from metal rods. The shafts will wear out the moving parts they rub against. To overcome this, "bearings" made from metal or plastic tubes are fixed in holes the shafts pass through. Bearings also reduce the amount of friction on a shaft, allowing it to move smoothly. You could replace all the shafts and bearings in the robot with cocktail sticks or thin wooden dowel, but it would not work so well or last so long.

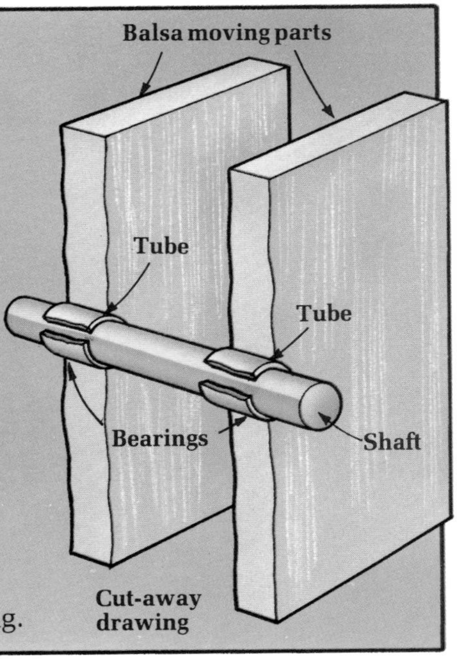

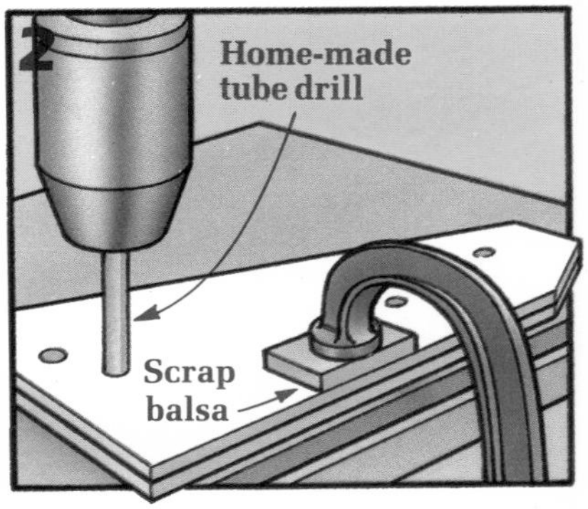

Gently clamp both shoulder sides together. Drill through the hole positions printed on the templates.

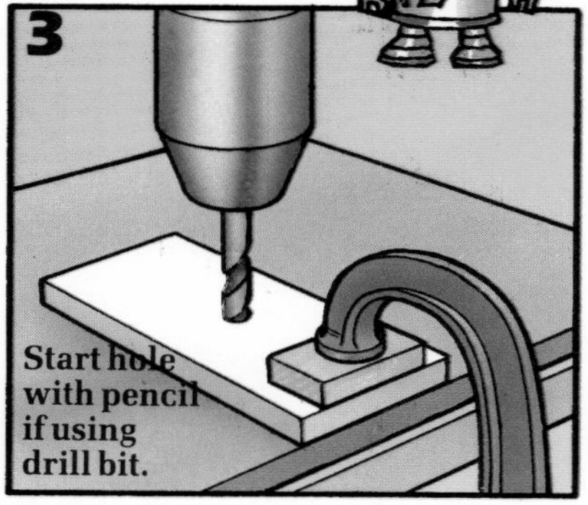

Gently clamp the shoulder top. Drill a 6mm hole where printed on the template. This hole is for the on/off light.

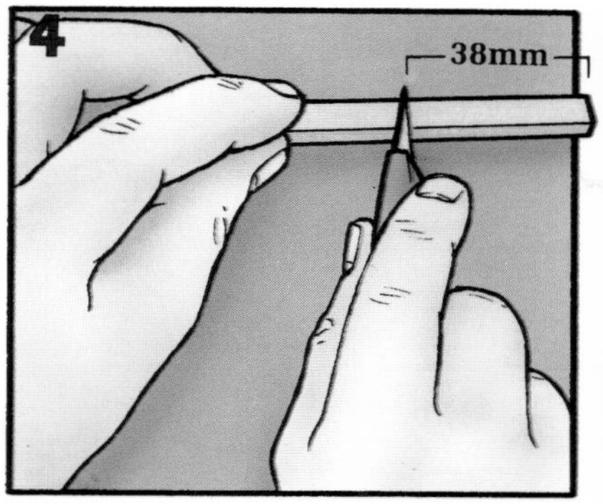

Mark a piece of 6mm spar 38mm long. Cut it with a tenon saw, or a craft knife by cutting half-way through each side.

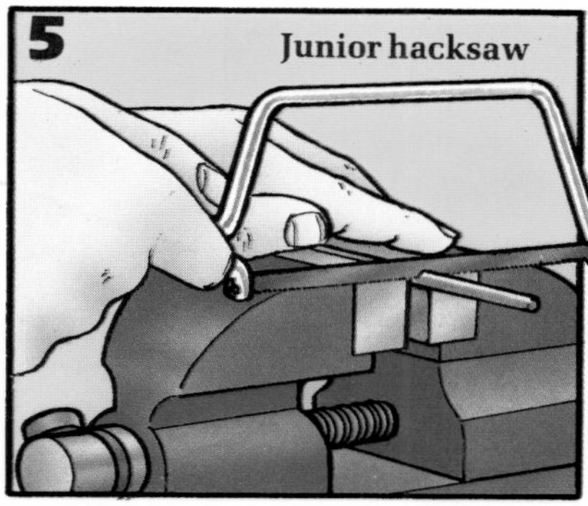

Gently clamp a length of metal tube as shown on page 8. Saw four 12mm and three 50mm pieces with a junior hacksaw.

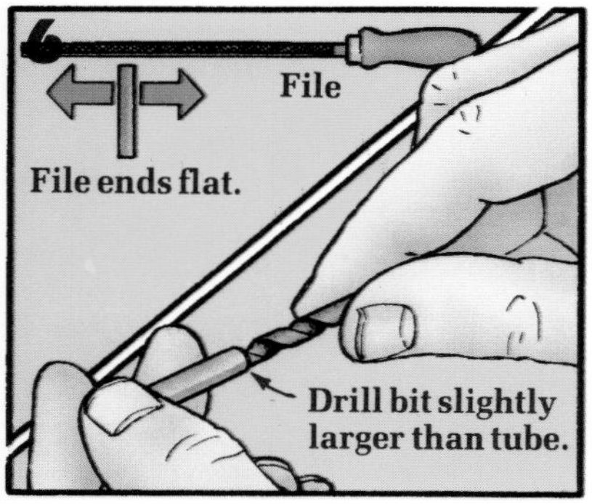

Hold the pieces of tube and file the ends flat. Twist a drill bit in the ends to remove rough edges, or burrs.

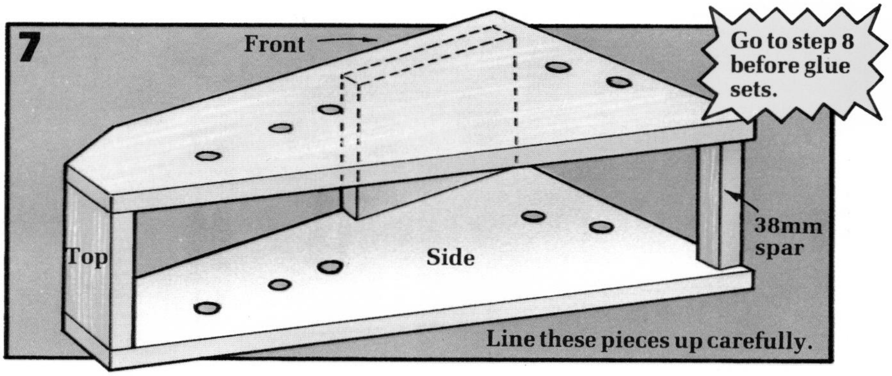

Glue together the shoulder sides, front, top and 6mm spar as shown above. The glueing positions are printed on the templates to help you line all the pieces up. Lightly sand the shoulder with fine-grade sandpaper and a block.

8 Test bench report

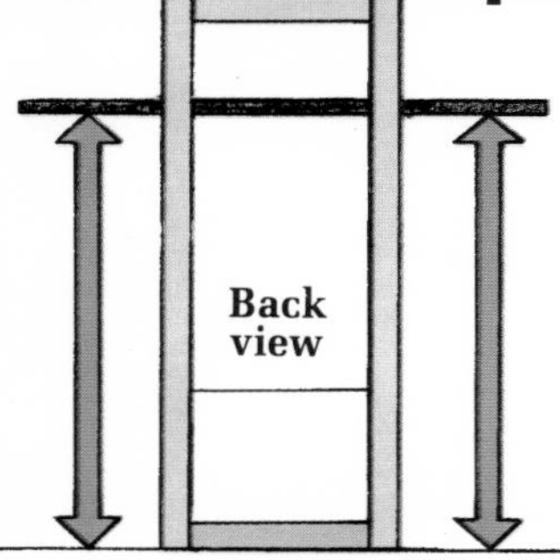

Shoulder

Push a length of tube through the bearing holes to check that they line up. If not, move the sides before the glue sets.

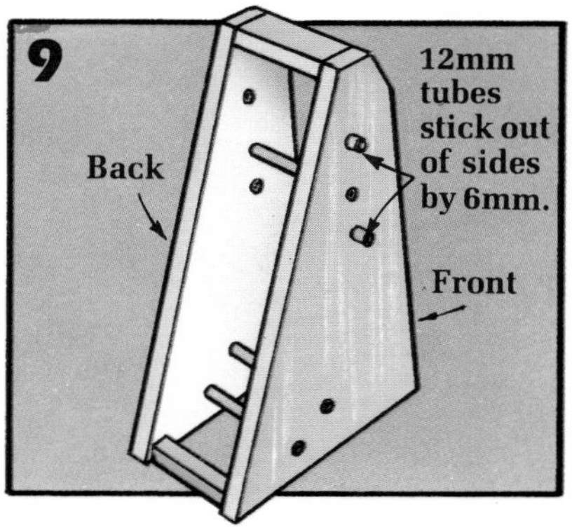

Push the pieces of tube into the holes in the shoulder sides. The tube lengths for each hole are on the templates.

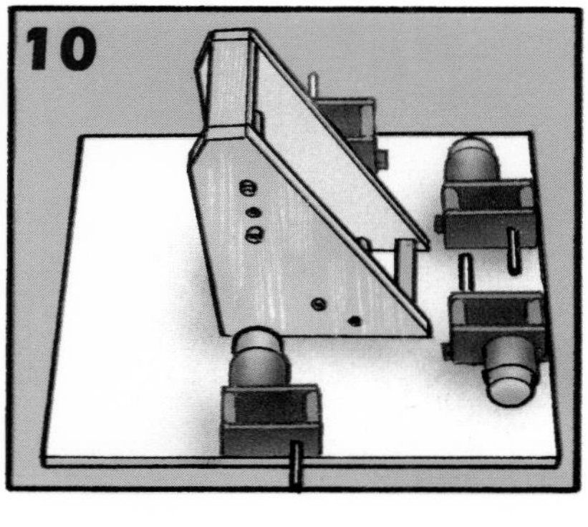

Glue the completed shoulder to the base over the glueing positions on the template, using plenty of glue.

Making the arm

You can find out here how to make the robot's arm (coloured dark blue on page 4). Some parts made at this stage are for attaching other parts of the robot later on, so don't worry if you cannot see what something is for just yet. Try to drill all the holes accurately.

Materials you need

Arm templates A, B, C, 1½m × 6mm balsa spar, 75mm × 3mm balsa sheet, metal or plastic tube (max. 6mm diameter), metal rod (to fit snugly inside tube), clear glue or balsa cement, paperclip. Tool kit including drill bit or home-made tube drill (same diameter as tube).

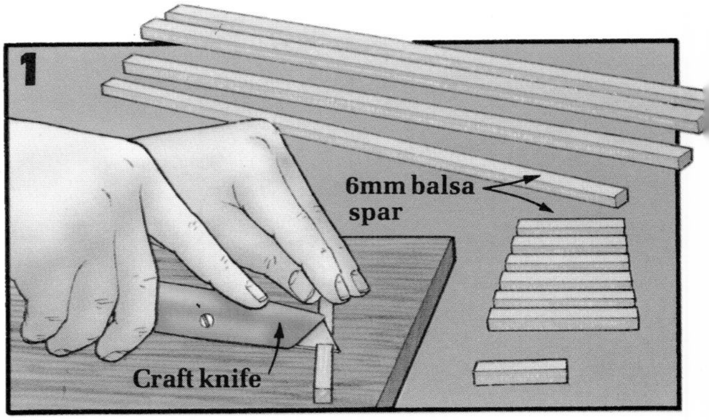

Measure and cut these lengths of 6mm spar: 4 at 250mm, 6 at 40mm, 2 at 24mm. Use a tenon saw, or a craft knife by cutting half-way through each side. Lightly sand the ends.

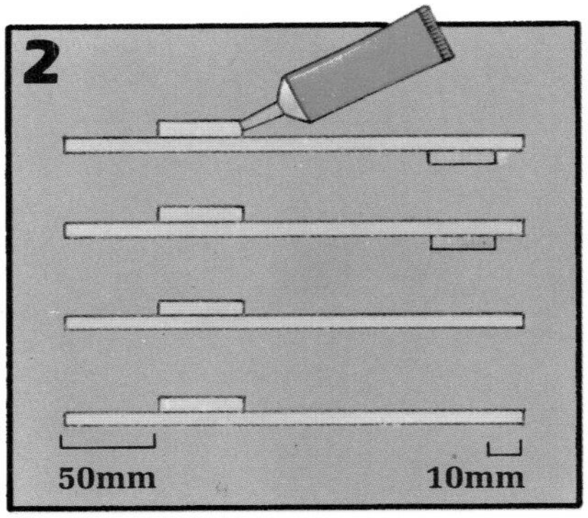

Glue the 40mm lengths of spar to the 250mm pieces in the positions shown above. Use plenty of glue for strength.

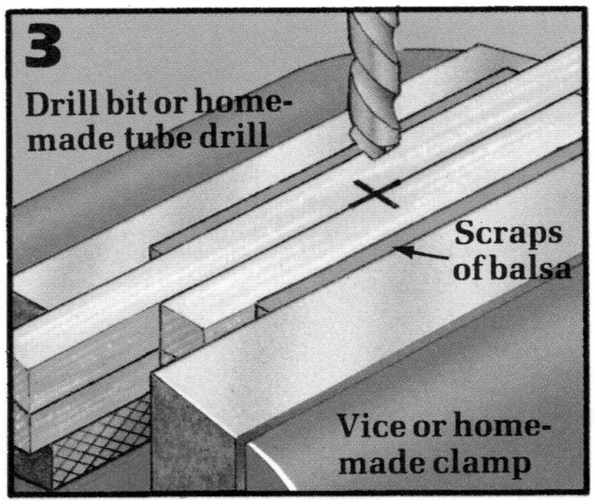

Measure and mark the centre of each joint. Drill a hole the same diameter as the tube, using a bit or tube-drill.

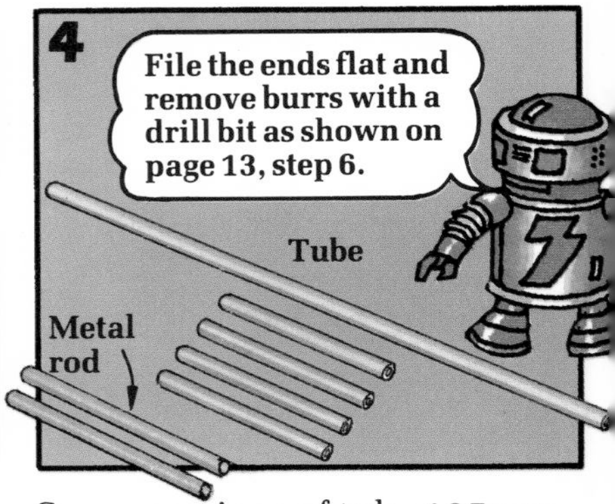

Saw one piece of tube 135mm long and four pieces 38mm long. Also saw two pieces of metal rod 70mm long.

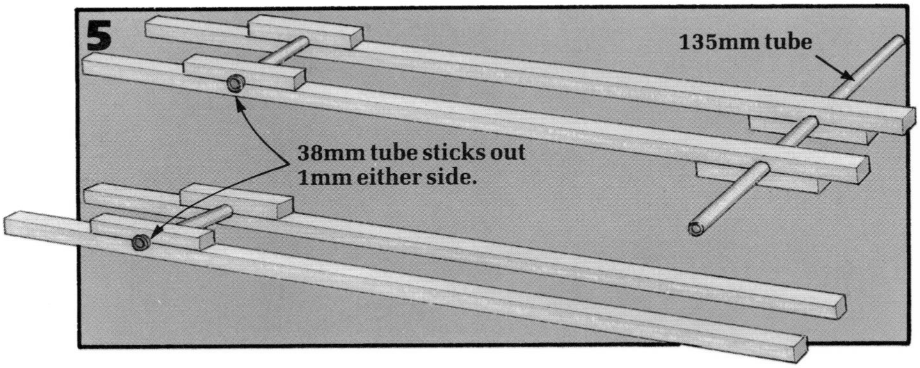

Push the 135mm tube and two of the 38mm pieces through the holes in the spars as shown above. Make sure the tubes are held firmly in the holes. If not, put a little glue in the holes to stick them in place, without getting glue in the tubes.

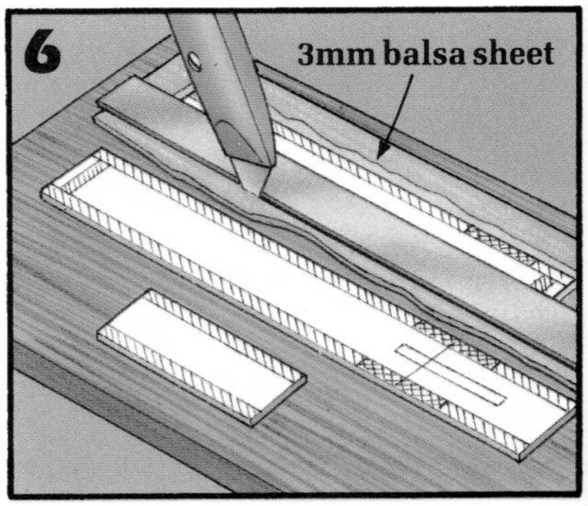

Glue arm templates A, B and C to 3mm balsa sheet. Cut round them, making sure you cut out the slot in piece A.

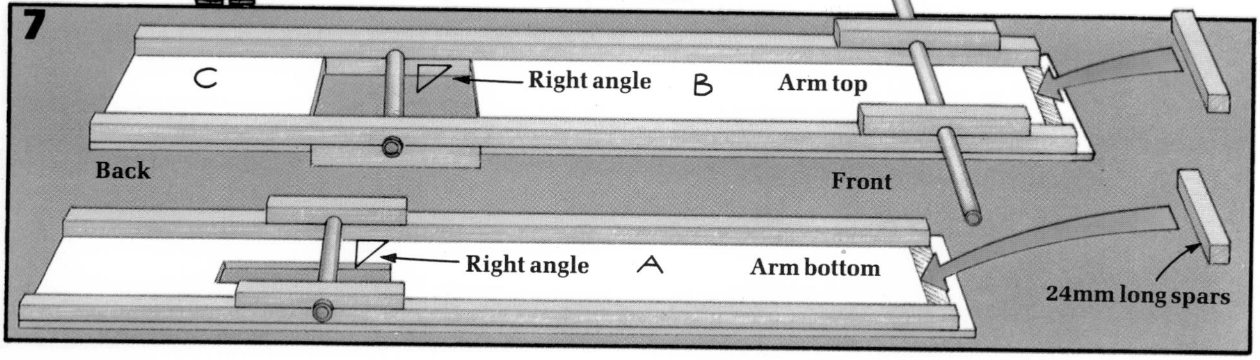

Carefully glue the parts made so far to the positions marked on the templates. Make sure the short lengths of tube are at right angles to the long spars. Make adjustments by moving the spars slightly before the glue sets. Then glue the 24mm lengths of spar on the template glueing positions, as shown by the arrows above.

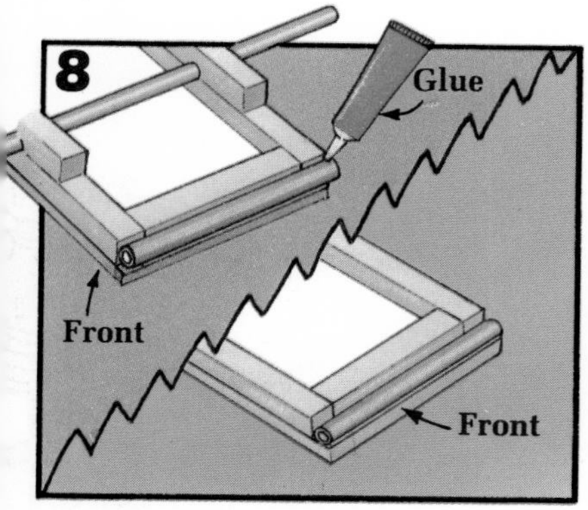

Glue the remaining 38mm lengths of tube in front of the 24mm lengths of spar. Use plenty of glue.

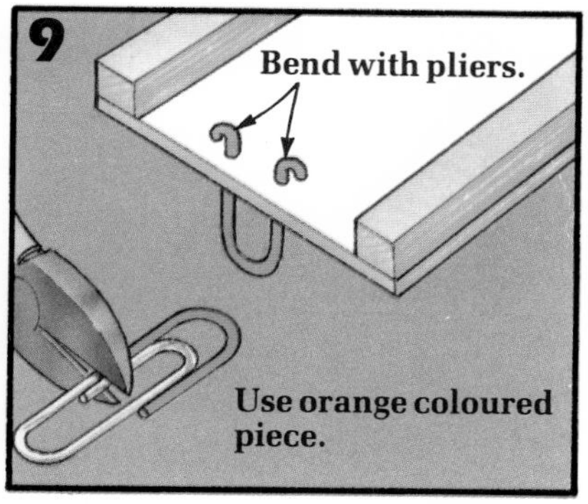

Cut the end off a paperclip with pliers or cutters. Push it through the marks at the end of template A and bend.

Test bench report

Robot arm

Position the arm bottom in the shoulder and slide one 70mm metal rod through the joint.

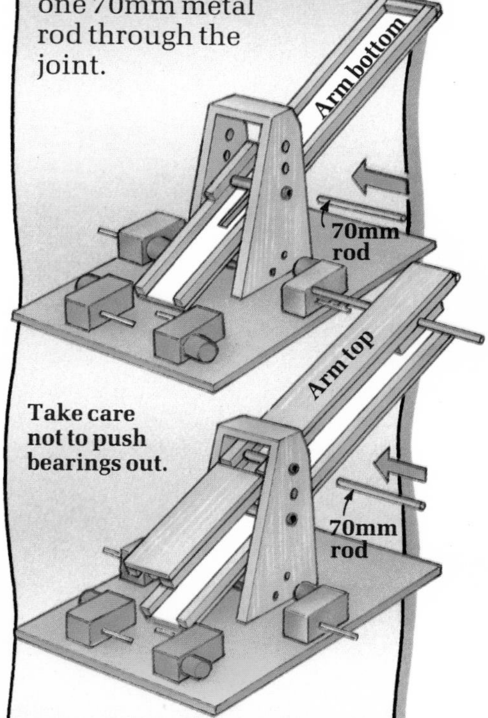

Repeat the same procedure with the arm top. Check both arm parts move freely up and down. Check the angles in step 7 if they do not.

Industrial arm robot

This industrial robot has three degrees of freedom. Its arm goes in and out

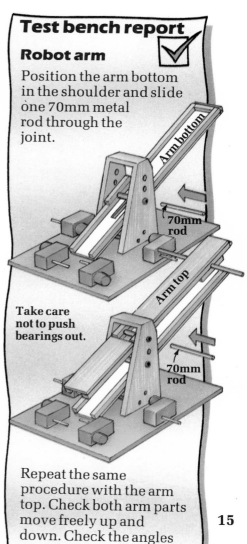

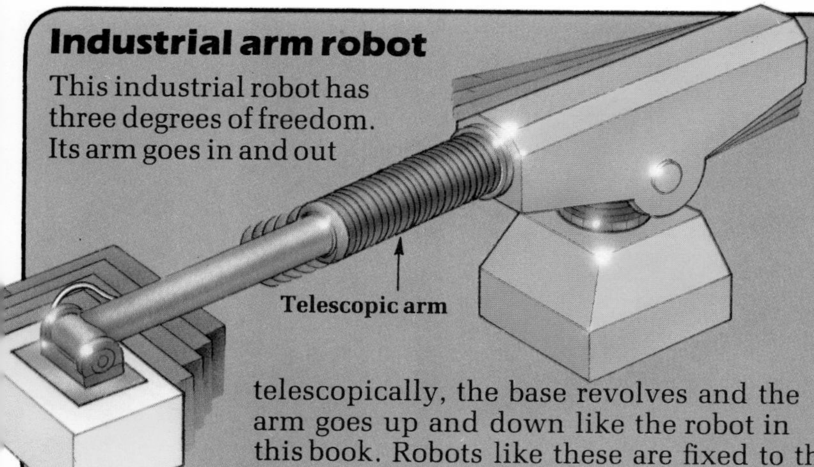

telescopically, the base revolves and the arm goes up and down like the robot in this book. Robots like these are fixed to the floor, so they cannot move around.

Making the gripper

The next four pages show you how to make the robot's gripper. The gripper is made up of two jaws. You need to repeat steps 6 to 9 to make both jaws. This part of the project is fiddly to make, so take your time and cut and drill all the parts as accurately as possible.

Materials you need

Gripper templates A to U, 3mm and 6mm thick balsa sheet, 6mm balsa spar, metal or plastic tube (max. diameter 6mm), metal rods (to fit snugly inside tube), stiff wire (paperclip or coathanger), fishing-line, two elastic bands (about 80mm long), two small tension springs or short elastic bands (max. stretch 25mm), battery for motors, sticky tape, clear glue.

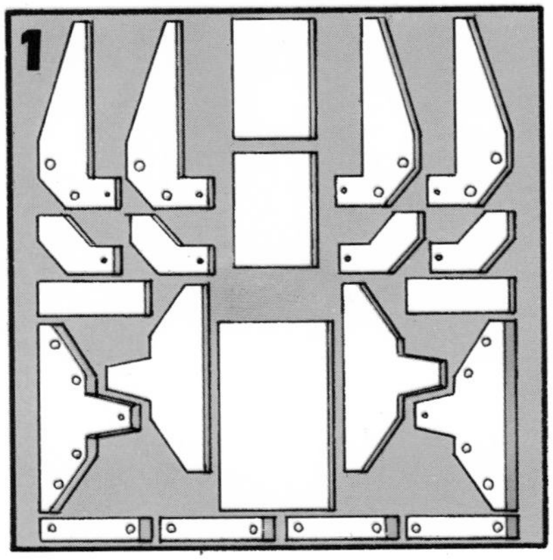

Cut out and glue gripper templates A to U to 3mm or 6mm balsa sheet according to the thickness printed on each template. Clamp matching parts together. Drill holes with a home-made tube drill or suitable bit, where printed on the templates. Also drill 2mm holes through parts G, H, I, J, K and N, where marked.

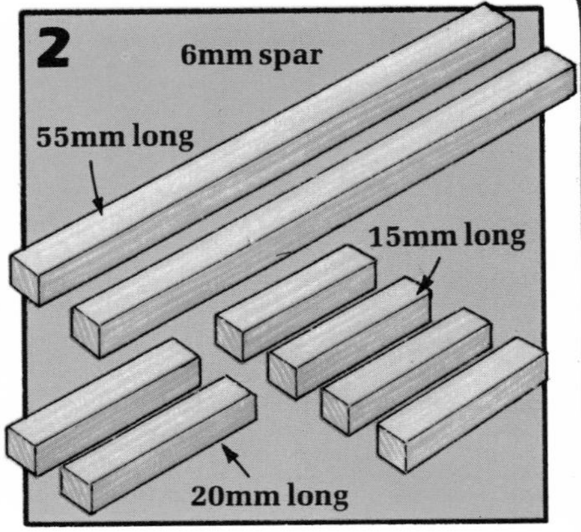

Measure and cut these pieces of 6mm spar: 2 at 55mm, 4 at 15mm and 2 at 20mm. Use a tenon saw or cut through the balsa both sides with a craft knife. Lightly sand the ends of each piece using fine sandpaper and a block.

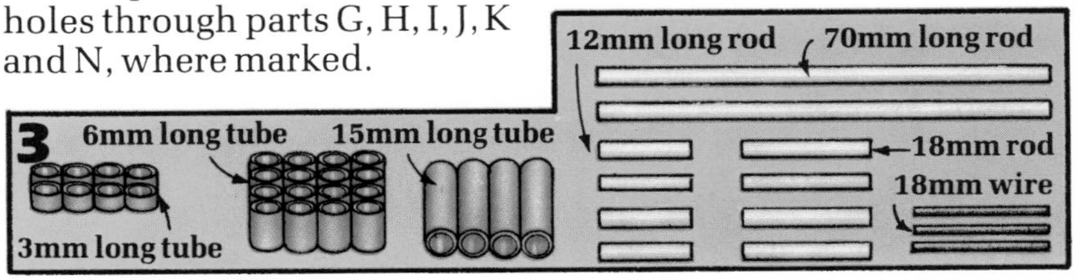

Cut the lengths of tube, metal rod and stiff wire as shown above. Use a vice or home-made clamp as shown on page 8.

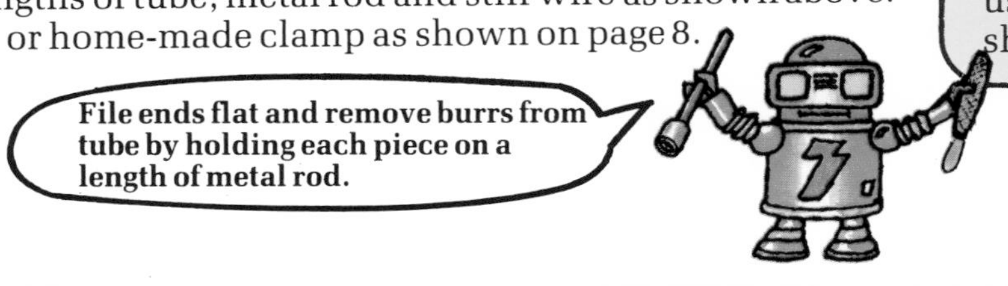

Robot wrists

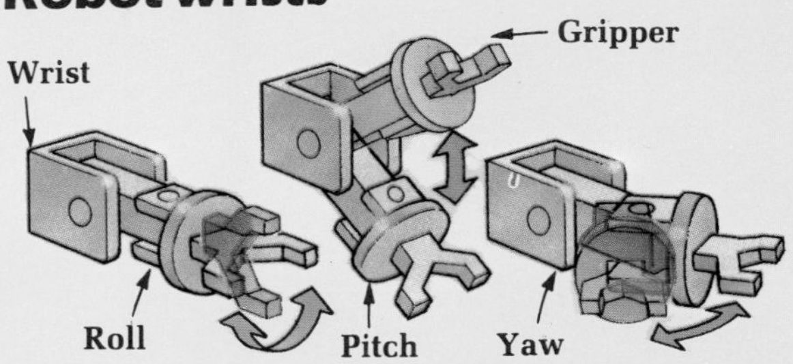

The part between a robot's arm and gripper is called the wrist. Wrists can be designed to allow three different movements, called yaw, pitch and roll. The wrist made in this book only pitches.

Grippers

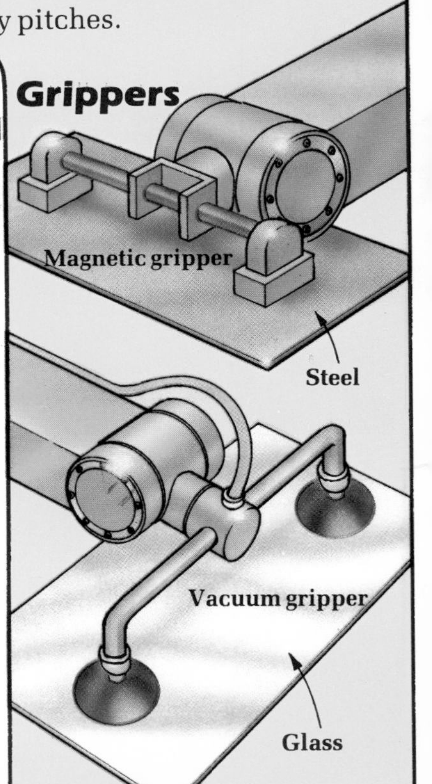

Industrial robots have grippers designed for the job they do. The pictures above show magnetic and vacuum grippers, often used to handle metal or sheets of glass.

4

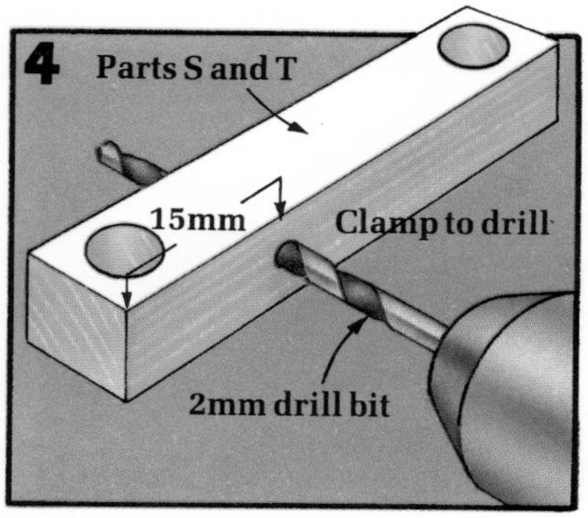

Drill a 2mm hole through the sides of parts S and T, 15mm from one end as shown. Clamp the parts to drill them.

5

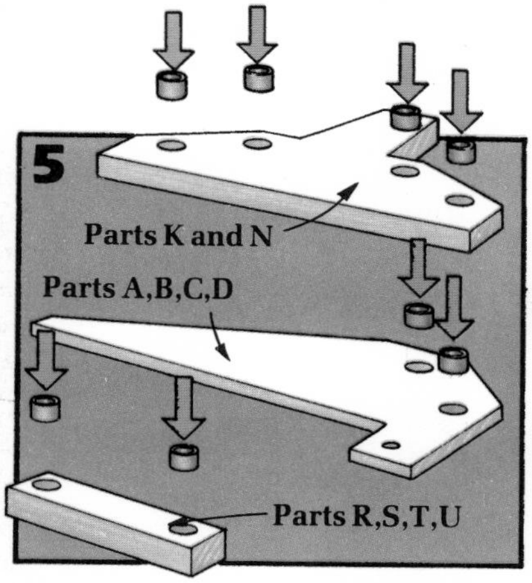

Push the 3mm and 6mm tubes through the holes. You could push a length of rod in first and slide the tube down.

6

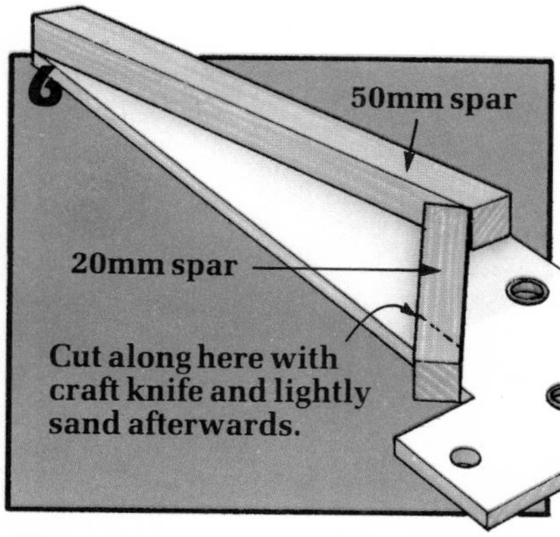

Glue a 55mm and 20mm spar to part C where printed on the template. Trim the end of the 20mm spar with a craft knife.

7

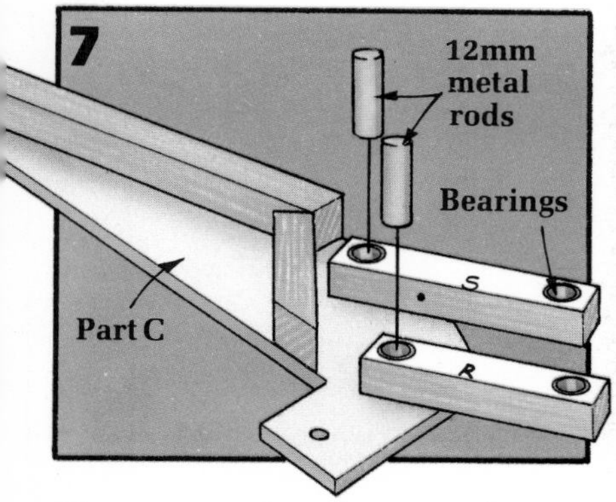

Place parts S and R over the holes in part C. Then slide two 12mm metal rods through the bearings as shown above.

8

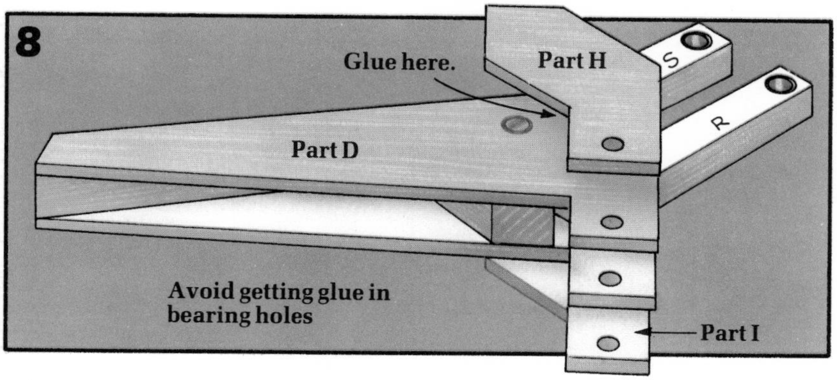

Glue part D to the top of the balsa spars so it fits over the metal rods and lines up with part C below. Then glue part I to part C and part H to part D, so that the shapes of all the parts line up. Avoid getting glue in the bearings.

9

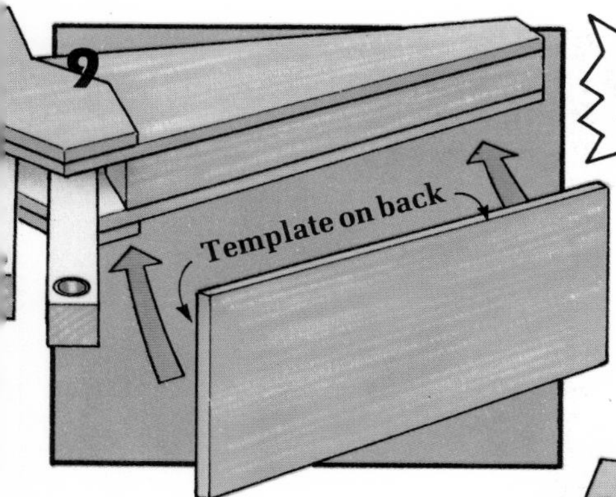

Apply glue to part E where printed on the template and stick it to the part made so far, as shown above.

Repeat steps 6 to 9 to make the other jaw.

10 Position the two jaws so that the holes in parts R, S, T and U line up with the holes in part K. Then carefully slide the four 18mm metal rods through the bearings in the holes. Next, push an 18mm length of stiff wire through the hole at the front.

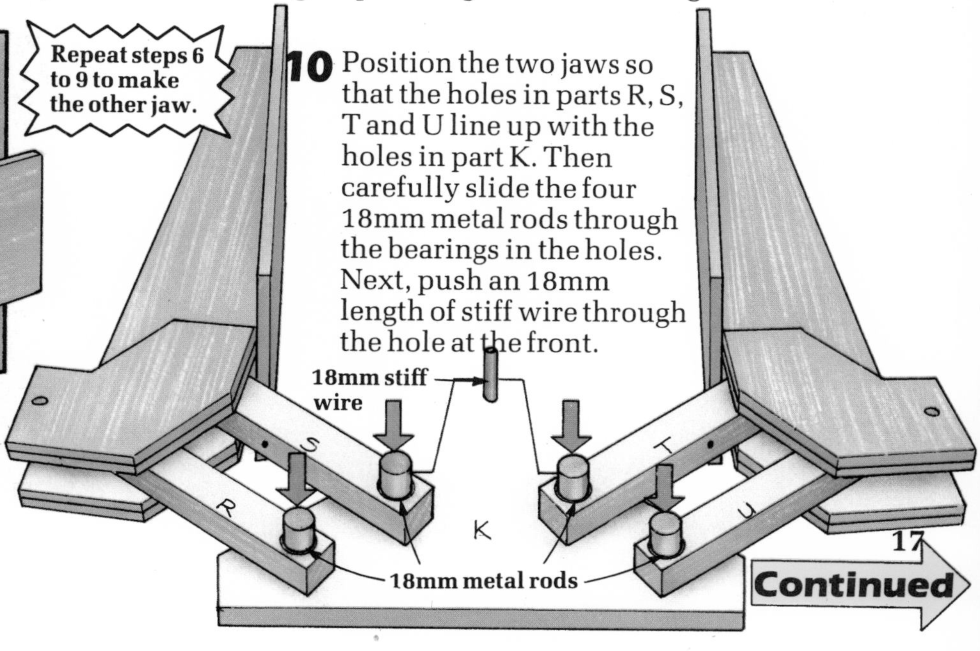

Continued

Making the gripper

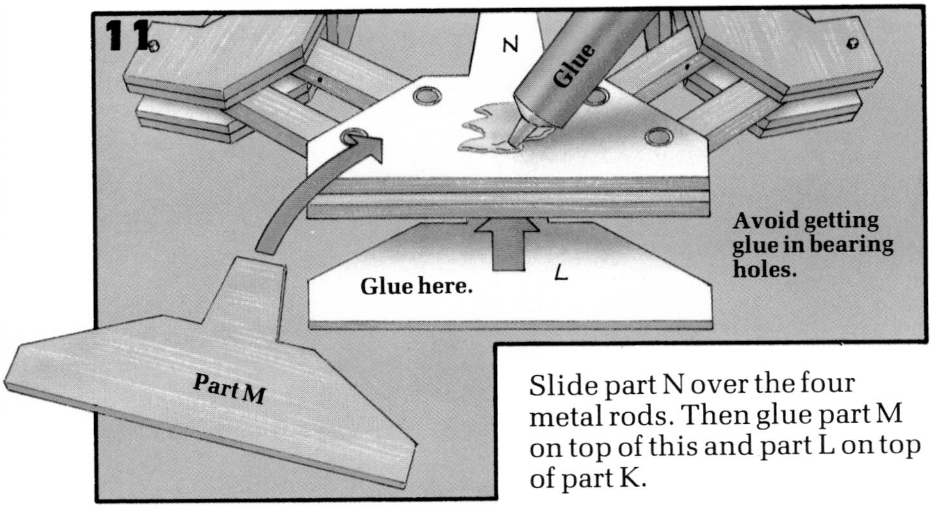

Slide part N over the four metal rods. Then glue part M on top of this and part L on top of part K.

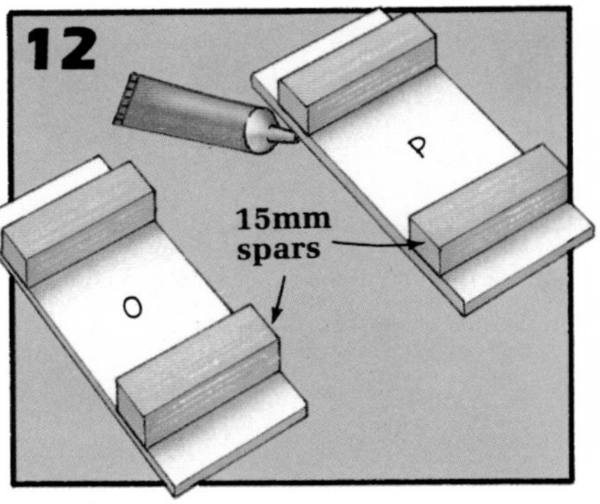

Glue four 15mm long spars onto parts O and P, making sure they line up with the positions on the templates.

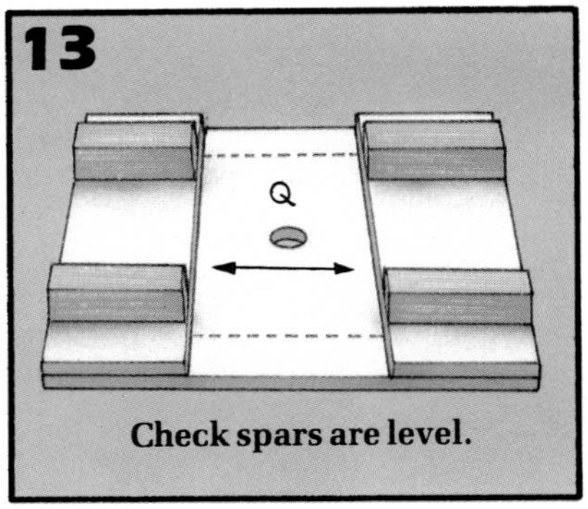

Now glue the parts just made onto the glueing positions on part Q. Make sure the spars are level. This part is the wrist.

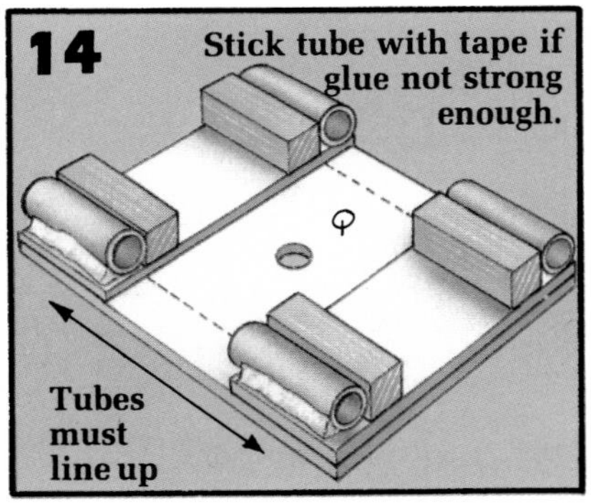

Glue four 15mm lengths of tube as shown, using plenty of glue. Try not to get any glue inside the tube.

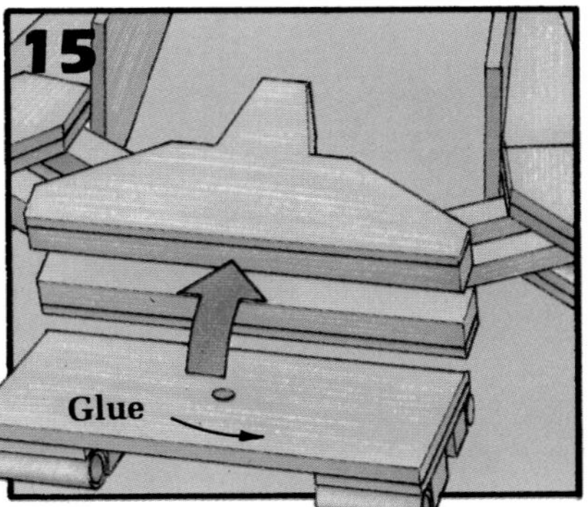

Glue the completed wrist to the gripper as shown, using the dotted lines printed on part Q as a guide to line it up.

Gripper and arm control lines

Fishing-lines wound round the shafts of the motors at the back of the robot operate the arm and gripper. These are called control lines.

1 Cut one length of fishing-line, or strong twine, 280mm long, one 400mm and one 500mm.

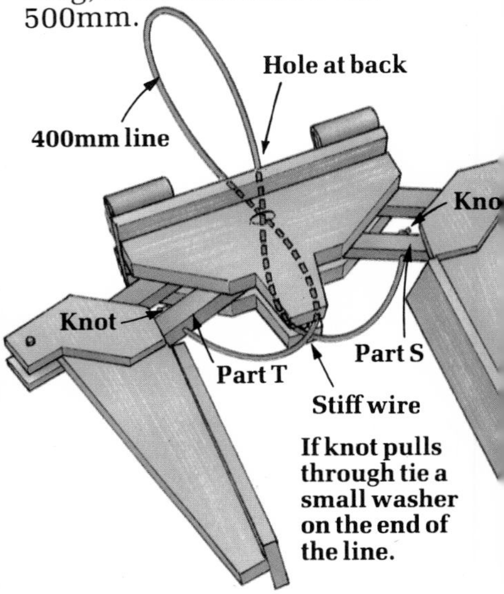

2 Thread the 400mm line through the hole at the back of the wrist, round the stiff wire and through parts S and T. Tie a knot at each end and tug to check it will not pull through the holes in S and T.

Test bench report

Gripper

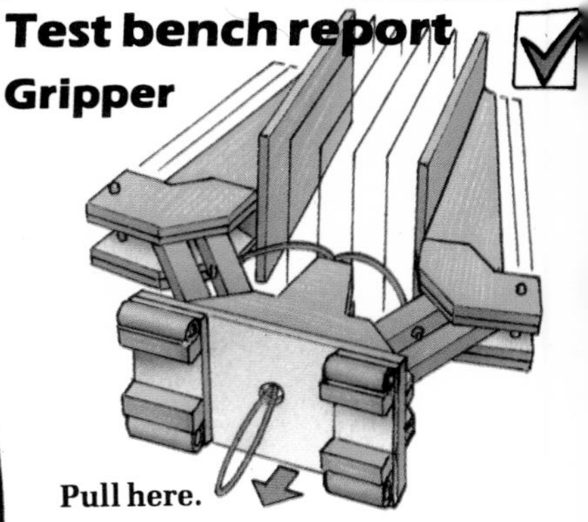

Pull on the line to close the jaws. Check the threading and knots if this does not work properly.

3 Remove the arm top (reverse of test procedure on page 15). Line up the tubes at the ends of the arm bottom and wrist. Then slide a 70mm metal rod through the tubes to connect both parts.

4 Hook a short tension spring or elastic band over the gripper line (shown red in the picture).

5 Tie the 50mm line (shown in blue) to the other end of the spring. Thread the line through the shoulder as shown.

6 Hook another spring over the wire loop. Tie the 280mm line (shown in red) to the spring and thread through the shoulder as shown.

7 Pull each line tightly round the motor shaft shown and tie it. Stick tape over the lines to stop them coming off.

8 Replace the arm top and join to the top of the wrist in the same way as for the bottom. Hook two elastic bands over the long tube and trap the other ends inside the jaws with the remaining pieces of stiff wire.

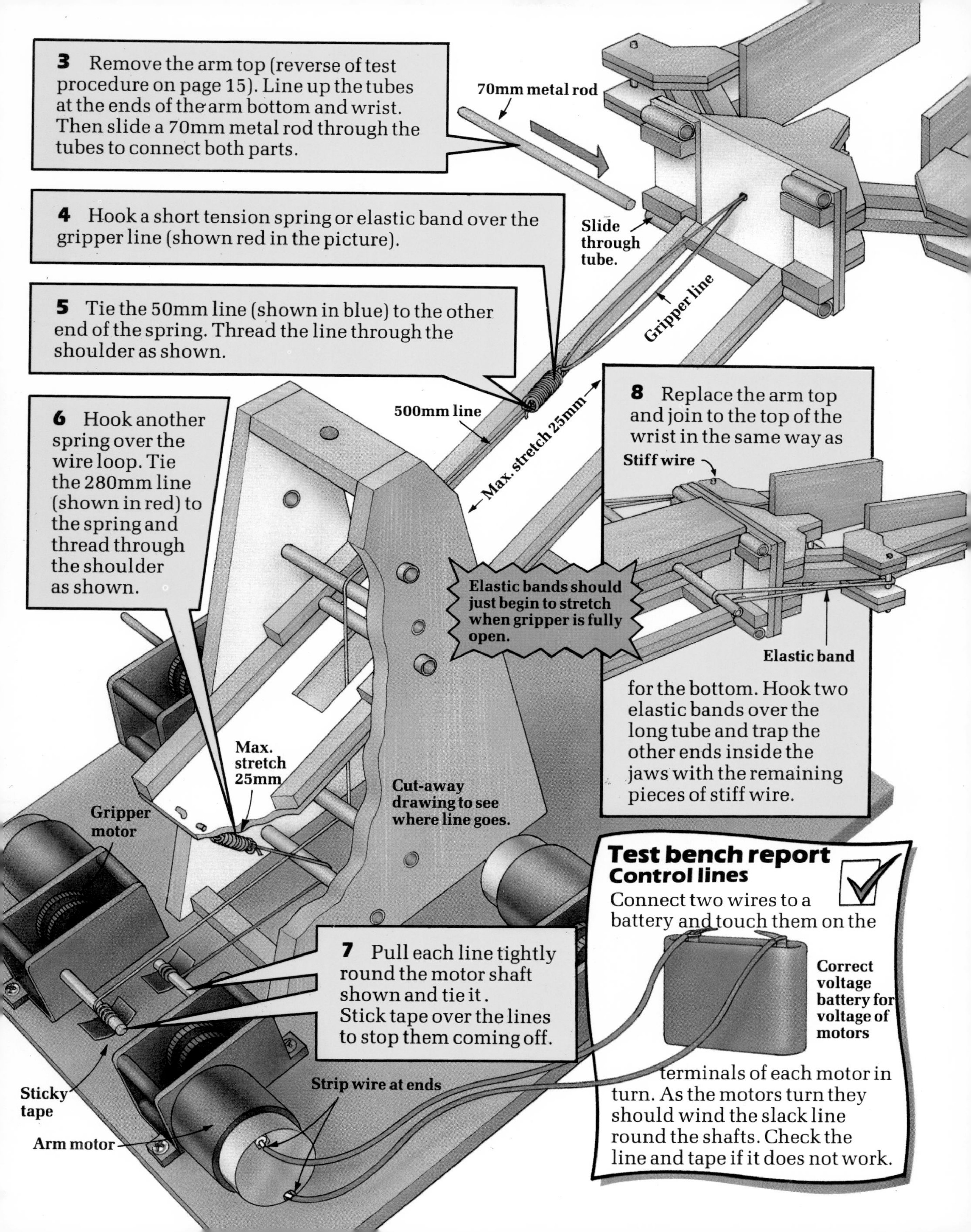

Test bench report
Control lines

Connect two wires to a battery and touch them on the terminals of each motor in turn. As the motors turn they should wind the slack line round the shafts. Check the line and tape if it does not work.

Customizing your robot

These pages show a cover to make, with tips on painting and ideas on customizing the different versions of the robot. There are lots of ways of making the robot look special by adding extra parts, such as a cardboard body. The cover has a flat top to make this easier to do.

Materials you need

Cover templates A to I, 1½mm balsa sheet, 6mm balsa spar, sticky tape, clear glue.

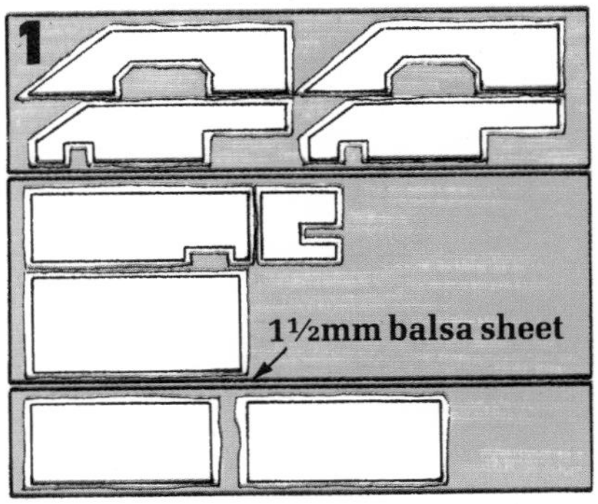

Cut out templates A to I and glue them to 1½mm balsa sheet. Cut round the templates with a craft knife.

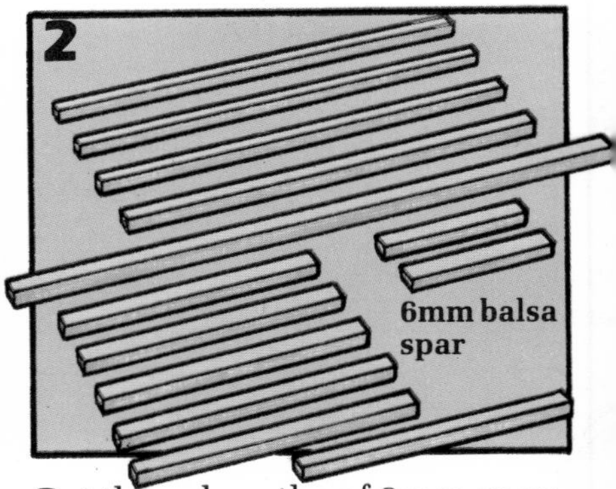

Cut these lengths of 6mm spar: 4 at 44mm, 6 at 70mm, 4 at 130mm, 1 at 159mm. Lightly sandpaper the ends.

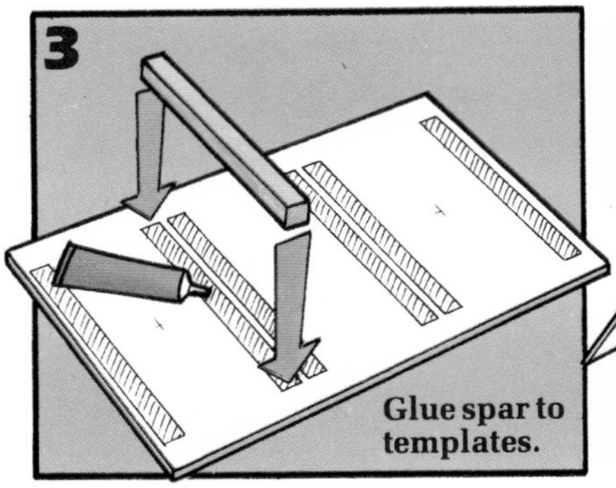

Glue the pieces of spar to parts A, D, E, F, G, H and I, on the glueing positions printed on the templates.

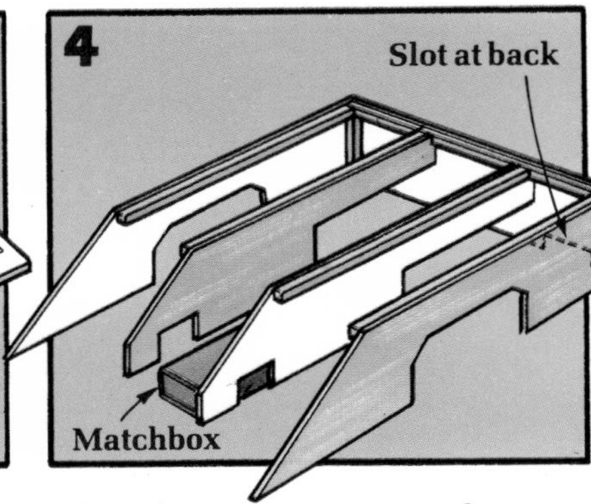

Glue the side pieces to the back as shown. You may have to support the sides with a matchbox while the glue sets.

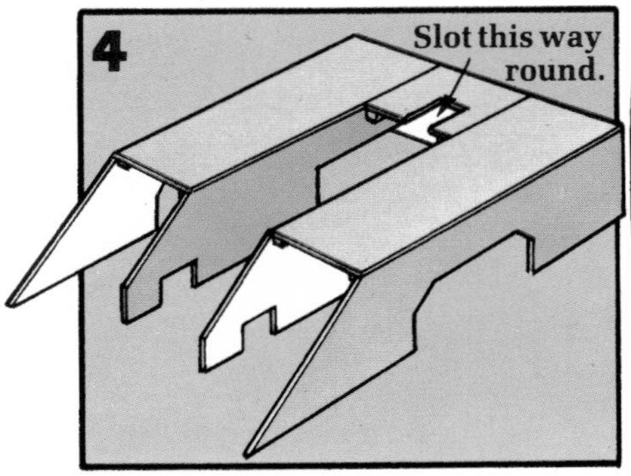

Glue parts B, C and E on top, template side down. Make sure the slot in part E faces the front of the cover.

Test bench report

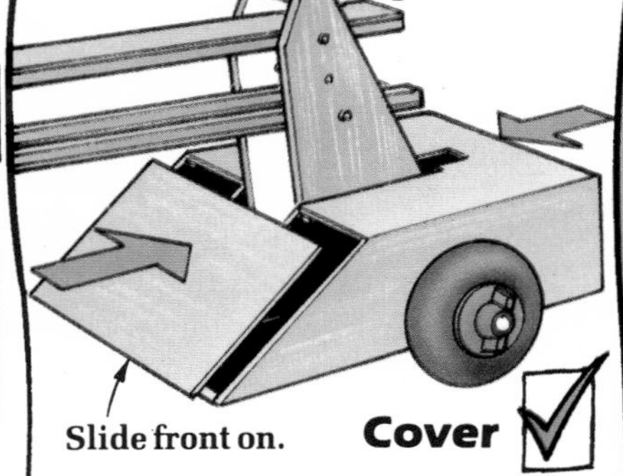

Slide the cover over the back of the robot and push part A onto the slanted sides. Check that the cover sits flat on the base. If not, tape it down but do not use glue.

Customizing ideas

These pictures show some customizing ideas. You could add stripes, numbers or mudguards to your robot, or even make it look like a bug-eyed science fiction monster.

Give your robot a name or number with rub-down lettering or paint brushed through a stencil.

The robot below is covered with synthetic fur fabric. Use clear glue to stick it on, but make sure moving parts are not obstructed.

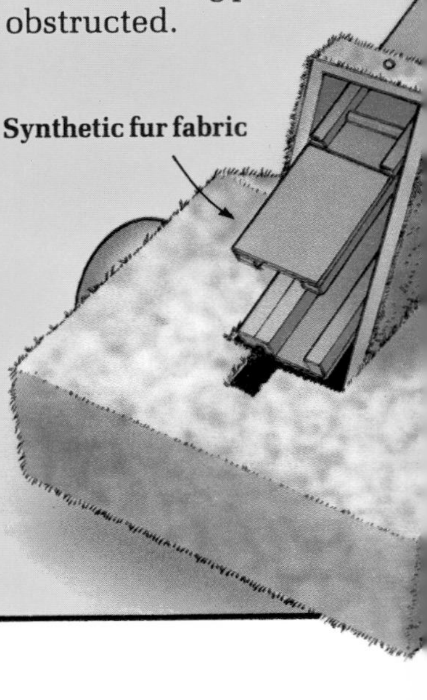

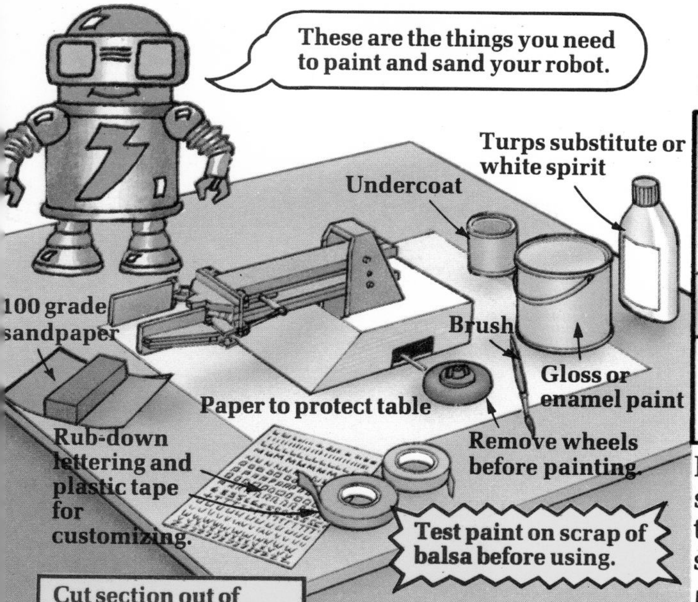
These are the things you need to paint and sand your robot.
Turps substitute or white spirit
Undercoat
100 grade sandpaper
Brush
Gloss or enamel paint
Paper to protect table
Rub-down lettering and plastic tape for customizing.
Remove wheels before painting.
Test paint on scrap of balsa before using.

Sanding

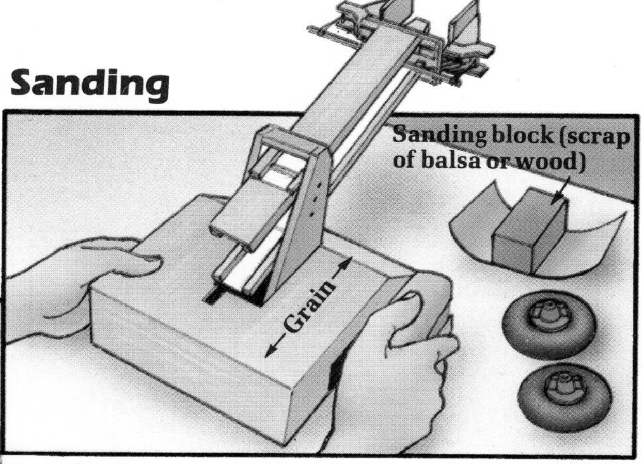
Sanding block (scrap of balsa or wood)
Grain

Lightly sandpaper the robot to give a smooth surface for painting. Gently rub the surfaces in the direction of the balsa grain with fine sandpaper wrapped round a block.

Painting

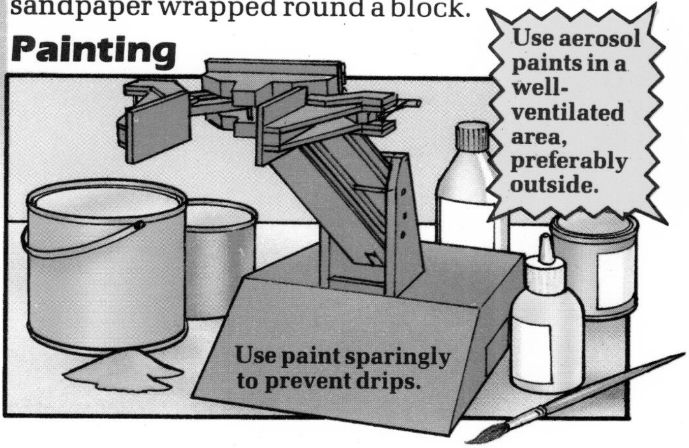
Use aerosol paints in a well-ventilated area, preferably outside.
Use paint sparingly to prevent drips.

Use an oil-based undercoat to seal the balsa and lightly sand again before painting a top coat of enamel paint. Avoid getting drips of paint in the joints.

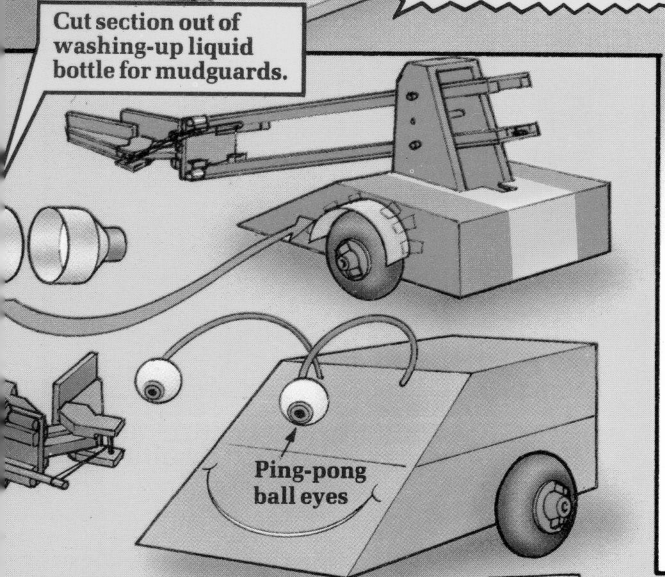
Cut section out of washing-up liquid bottle for mudguards.
Ping-pong ball eyes

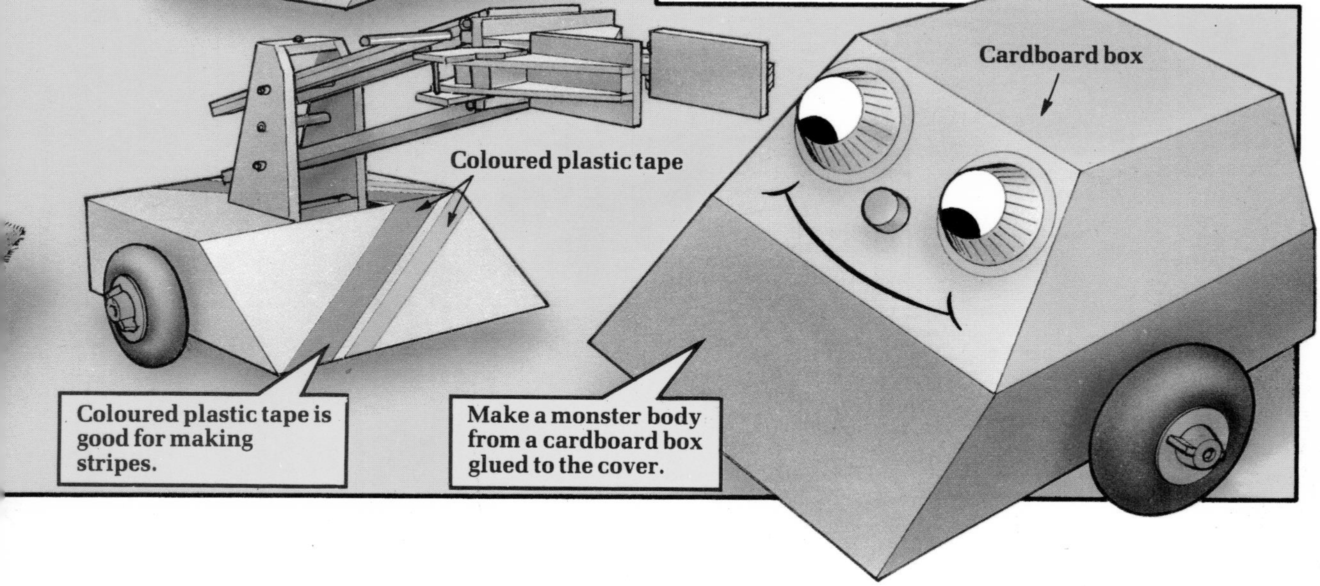
Cardboard box
Coloured plastic tape
Coloured plastic tape is good for making stripes.
Make a monster body from a cardboard box glued to the cover.

Electronics and soldering

The next few pages explain the electronic parts needed to control the robot with a home computer. Here you can find out about electronic circuits and soldering. Electronics is about the control of tiny electric currents with devices called components, soldered together to make circuits.

Things for soldering

Soldering is a way of joining two bits of metal together with another metal called solder. These are the things you need.

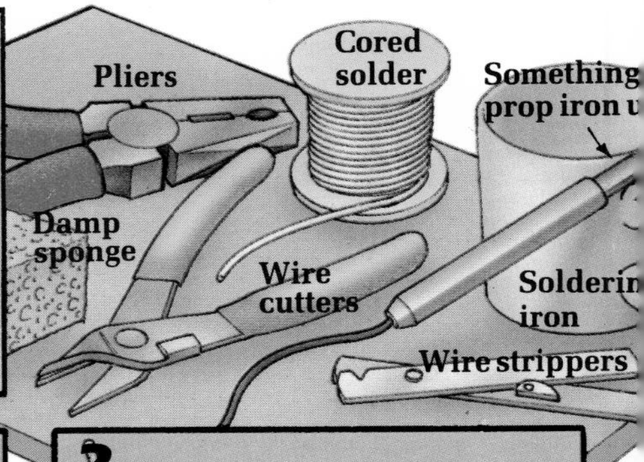

What is a circuit?

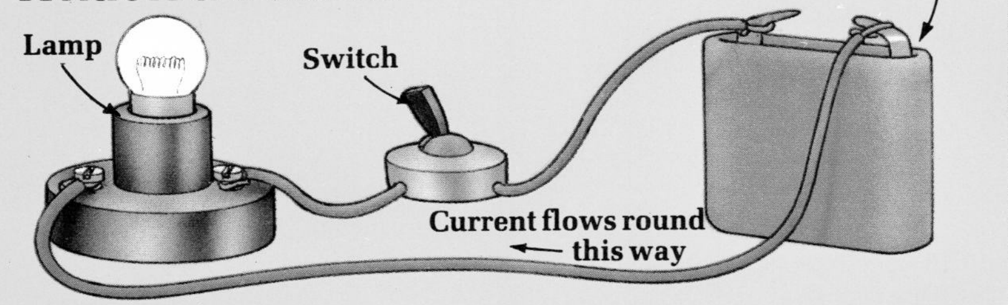

A circuit is made by connecting components to a battery so electricity flows through them to get a certain effect. The picture above shows a simple circuit. When the switch is on, current flows from the battery, through the lamp, through the switch and back to the battery.

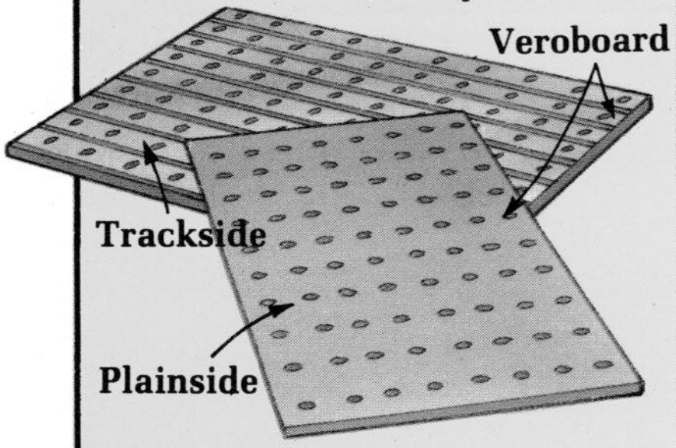

You connect components by soldering them to a special board called Veroboard. This has rows of holes with copper tracks on the back linking them together. Component legs are pushed through the holes and soldered to the track. Current flows along these tracks.

1 How to solder

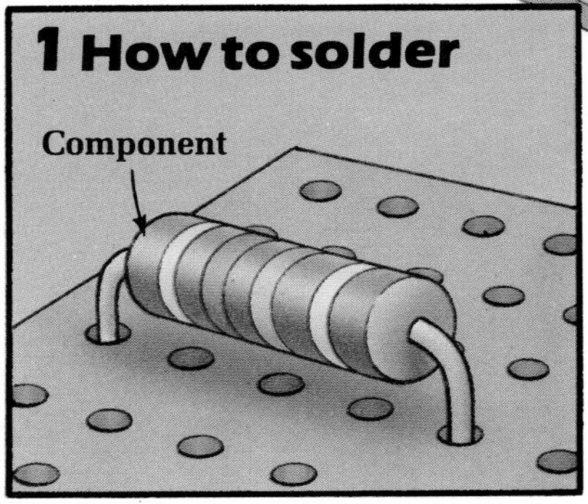

Bend the component's legs and push them through the holes in the plainside of the Veroboard.

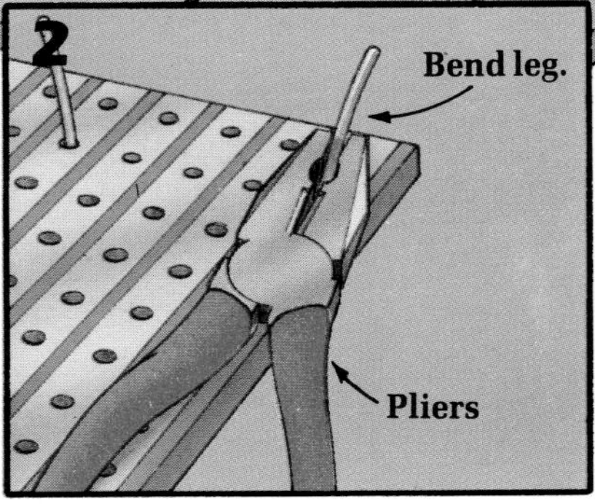

Turn the Veroboard over and bend the legs out slightly using pliers. This stops the component falling out.

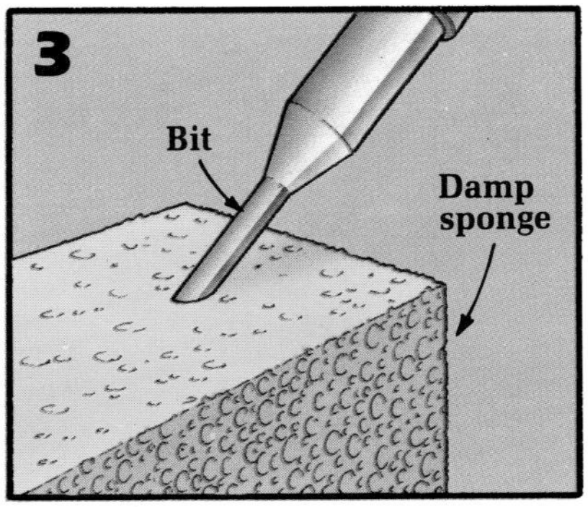

Plug in the soldering iron and wait until the bit heats up. Then wipe the bit on a damp sponge to remove old solder.

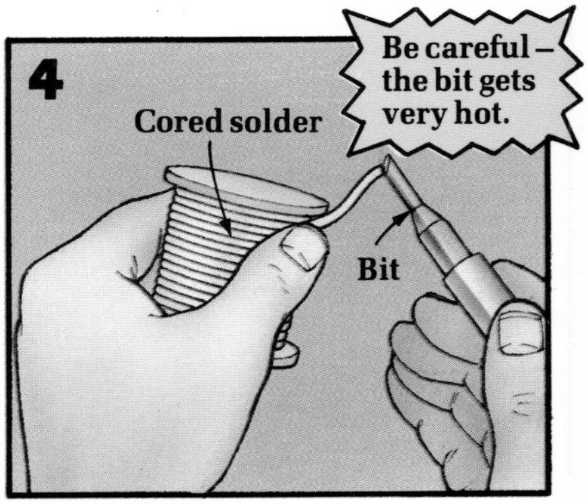

Carefully touch the bit with solder so that a drop clings to it. This is called "wetting" the bit.

5

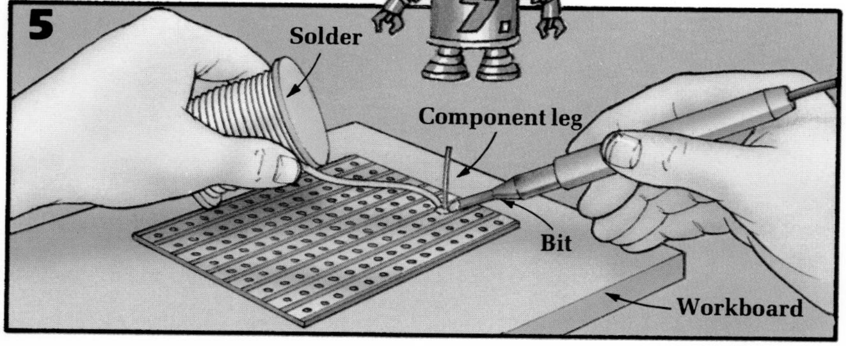

Touch one side of the leg, where it touches the track with the bit, and at the same time touch the solder on the other side. Hold them there for about a second, until a blob of solder flows round the leg. Then let the joint cool.

6

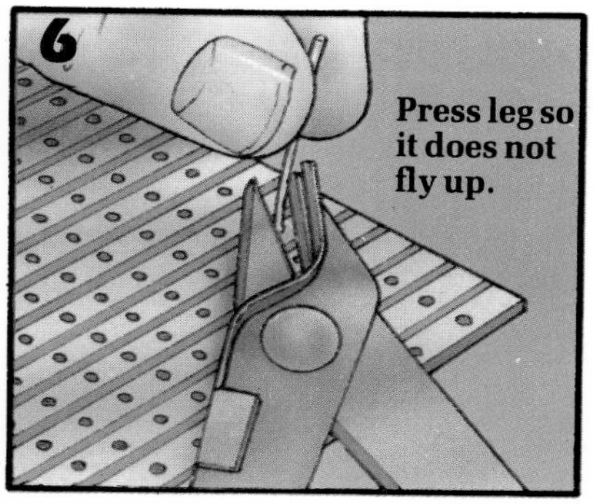

Trim the legs with wire cutters close to the solder. Hold the board away from your face and put your finger on the leg.

Stripping wire

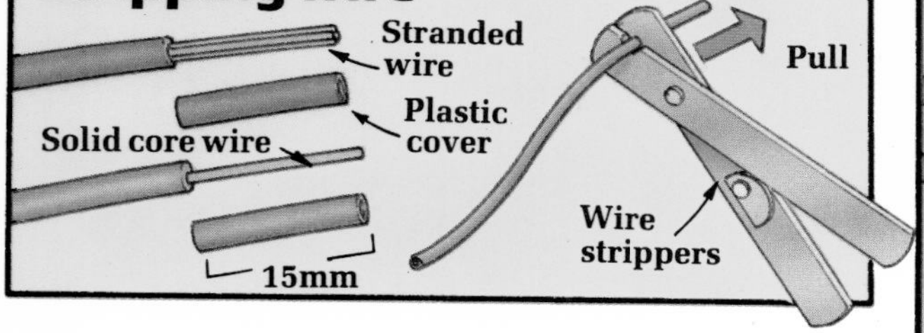

Wire is used to connect one track to another in a circuit, and to connect to a battery. Remove about 15mm of plastic from each end, using wire strippers adjusted to cut only the plastic, not the metal core. Grip the strippers firmly, and pull while holding the wire.

Tinning wire

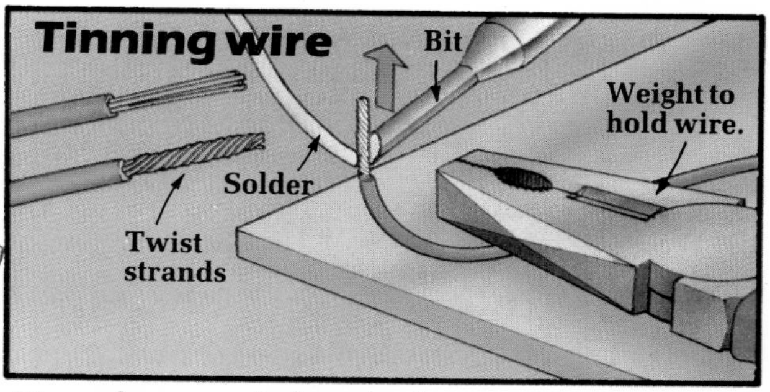

To make a good electrical connection and stop stranded wires coming apart, you need to coat the stripped ends with solder. This is called "tinning". Stroke the wire with the bit and solder until lightly coated.

Practise your soldering

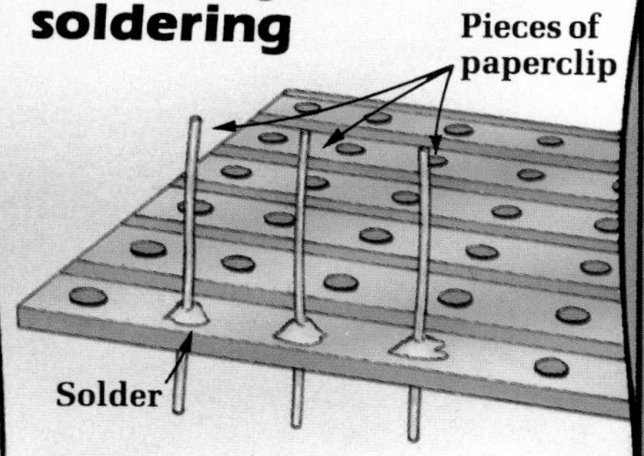

Score a small piece of Veroboard with a craft knife, and snap it to break a piece off. Try making lots of joints in adjacent holes using pieces of straightened paperclip.

Solder check

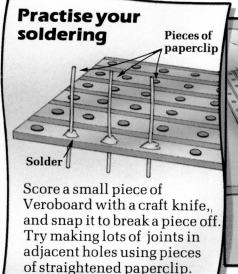

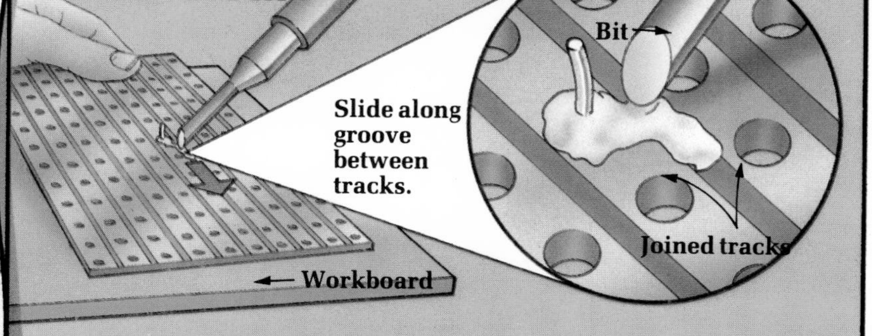

It is very important not to join the tracks with solder accidentally. Inspect your practice soldering to see if this has happened. If so, remove any solder by carefully running a hot bit between the tracks.

Always pull the plug out when you have finished soldering.

Electronic components

These pages show all the components used to build the switching circuit and sensors, with hints and tips on identifying them. The ones you buy may not look exactly the same. If you cannot identify a component or its legs, ask your supplier to help.*

Resistors

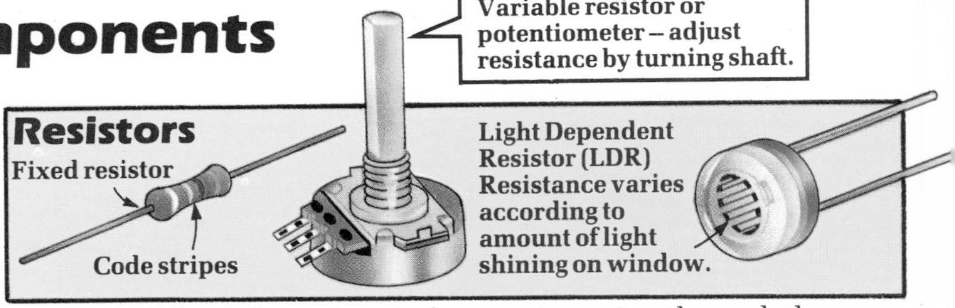

Resistors reduce the amount of current passing through them. Some do this by a fixed amount and others are adjustable. Coloured stripes on fixed resistors are a code to tell how strong their resistance is. Crack the code using the chart opposite.

Diodes

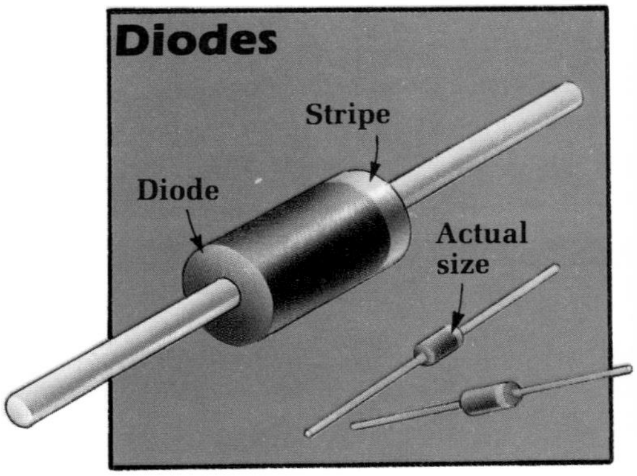

Current flows in one direction through diodes – like a one-way street for electricity. They have a stripe at one end which identifies which way round you connect a diode in a circuit. Current only flows when a diode is correctly connected.

LEDs

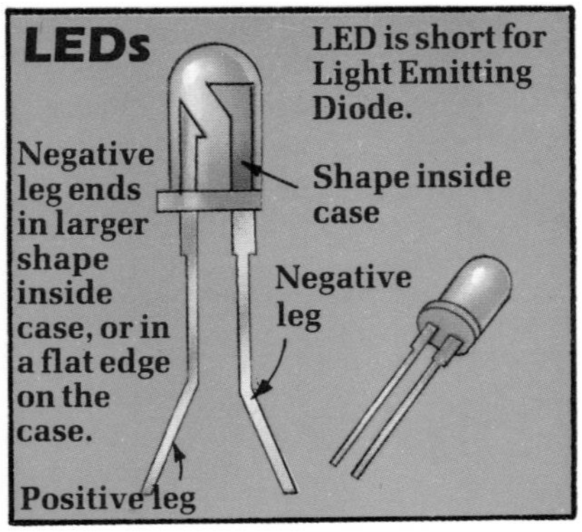

LEDs glow like tiny bulbs when current passes through them. Like ordinary diodes, current only goes one way. They have a positive leg, connected to the positive terminal (+ve) of a battery, and a negative leg connected to the negative terminal (−ve).

Transistors

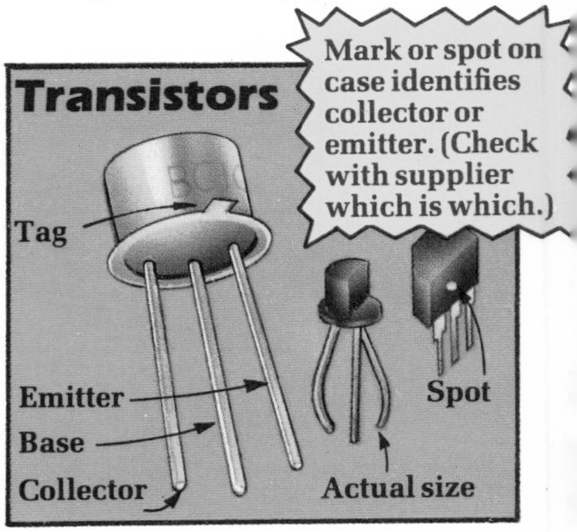

Transistors act like switches to turn current on or off, or to control the strength of current. They have three legs, called "base", "collector" and "emitter". The centre leg is usually the base and the case has a mark to identify the others.

Relays

A relay is a special kind of switch activated by an electromagnet. The picture on the right shows a cut-away of the type used in the project. The coil in the centre becomes an electromagnet when current is passed through it. This attracts the springy arm above it, switching it from one of the contacts at the end to the other. The magnet works as long as current flows, but as soon as it is turned off the arm flicks back to the other contact. This effect is used to switch the robot's motors on and off. There are many types of relays, some with more than one switch inside. See page 40 to find out which you need.

Relay shown actual size.

These are called single-pole (or switch) changeover relays. Only some makes have a clear plastic cover.

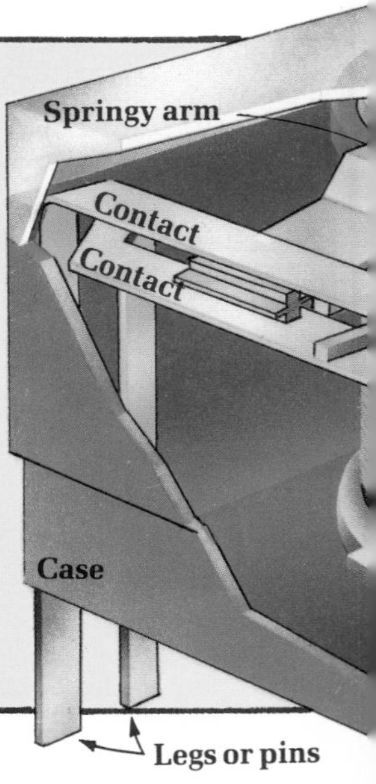

*It is a good idea to take this book with you to the shop. See the shopping list on page 40 for the components you need.

Resistor codes

Colour		Digit or number of 0s
	Black	0
	Brown	1
	Red	2
	Orange	3
	Yellow	4
	Green	5
	Blue	6
	Violet	7
	Grey	8
	White	9

How many ohms are these resistors? Answers on page 48.

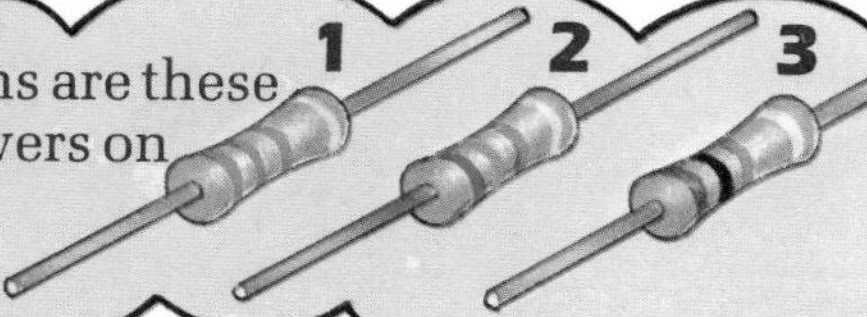

This shows how to read resistor codes.

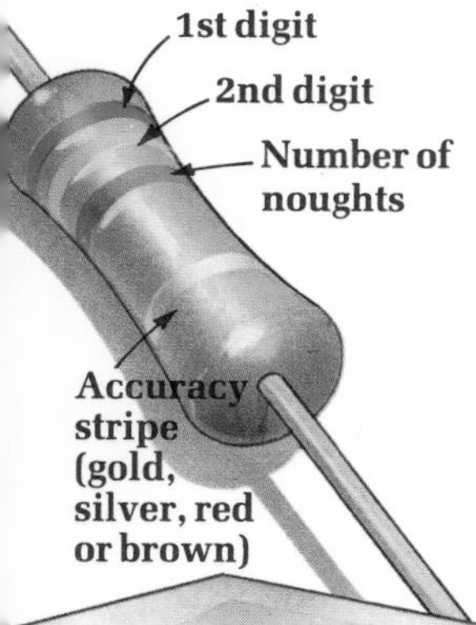

Resistance is measured in ohms (written Ω or KΩ for 1,000 ohms). One stripe shows how accurate the resistor is (don't worry about this one), and the others show its resistance. The first and second stripes give the first two digits in the number of ohms. The colour of the third shows how many noughts to add to this.

See page 41 to find out how to identify relay pins.

Sensors

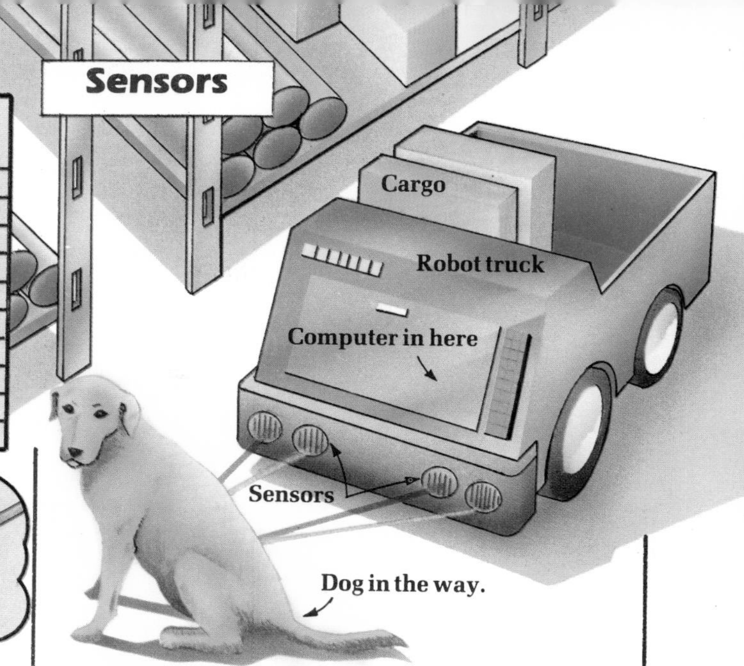

Robots need electronic senses, or sensors, to give their computers information, or feedback, about the outside world or the robot itself. This robot truck has a kind of radar sensor to detect obstacles in its path. The instant anything is detected it sends a message to an on-board computer, which is programmed to control the robot to steer clear.

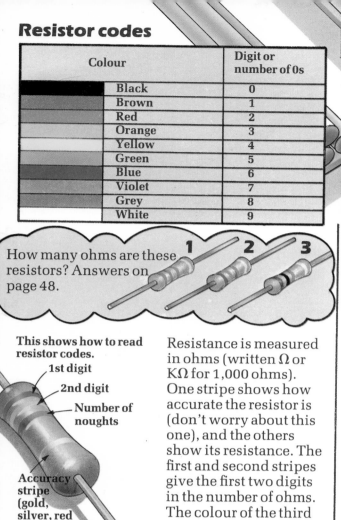

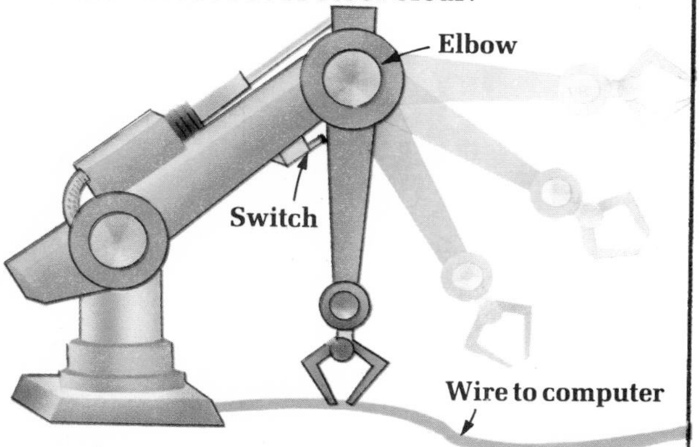

The simplest kind of sensor is a switch. The arm robot above has a switch like a lamp switch under its elbow. The switch is connected to its computer. As the arm bends, it touches the switch. This triggers the computer to turn off or reverse the arm motor to avoid damaging the robot. The project arm and gripper have similar sensors. A mobile robot could also have a switch under its bumper, to detect collisions.

Making robot sensors

The robot's sensors tell its computer when the arm is either up or down and whether the gripper is open, closed or holding something. These are switches, made from two pieces of metal which make contact when the arm or gripper have moved as far as they will go. The computer program on page 36 makes the motors stop or change direction when the switches make contact.

Materials you need

3m bell wire or ribbon cable, 2 × 50mm stiff wire (straighten out a couple of paperclips), insulation or sticky tape, double-sided tape, 50mm long steel nail or metal tube, 2 compression springs about 5mm diameter × 25mm long (e.g. old ballpoint pen springs) drill bit same diameter as springs, brass shim or tin foil about 60mm square, thin foam rubber about 60mm square × 5mm thick, glue, drawing pin, 12mm or 6mm balsa spar, soldering and tool kits.

1 Arm down sensor

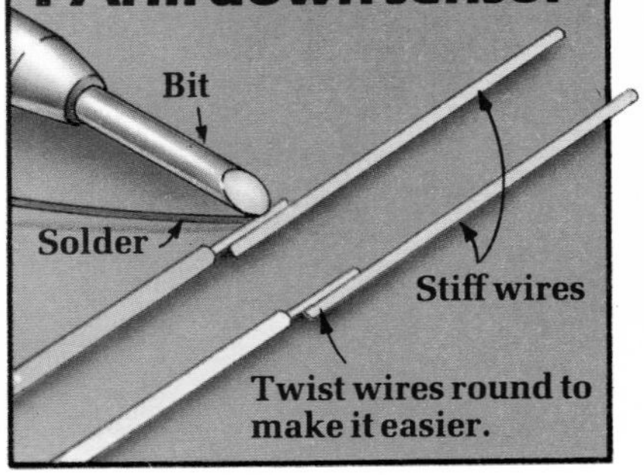

Measure and cut two 400mm long wires and strip and tin both ends. Solder one end of each to two 50mm long bits of stiff wire.*

2

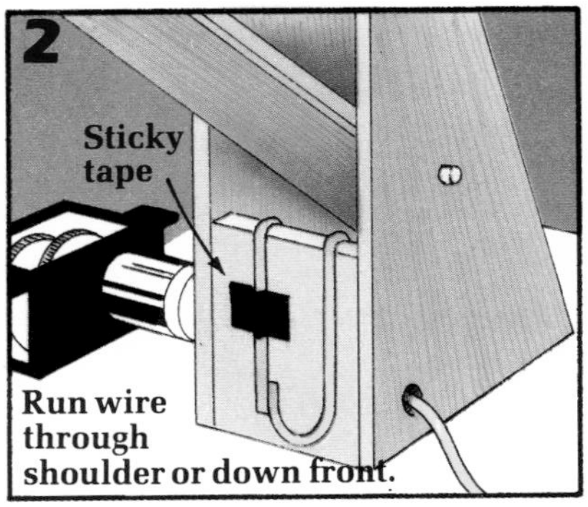

Bend one of the stiff wires over the front of the shoulder as shown. Cut off a piece of tape and stick the wire to the front of the shoulder.

3

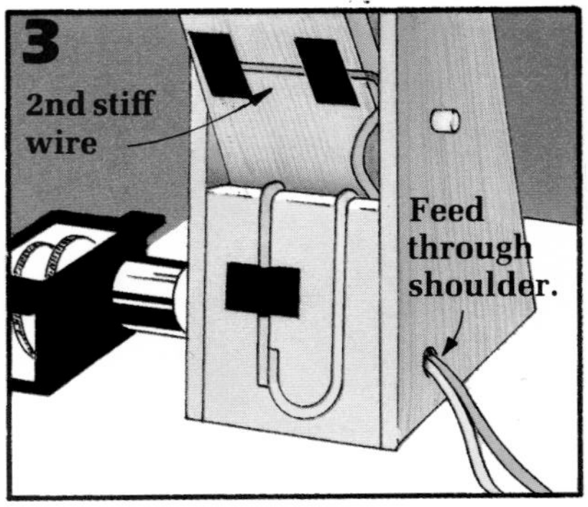

Stick the other stiff wire under the arm with tape. Position it so that the stiff wires touch each other when the arm is right down.

1 Arm up sensor

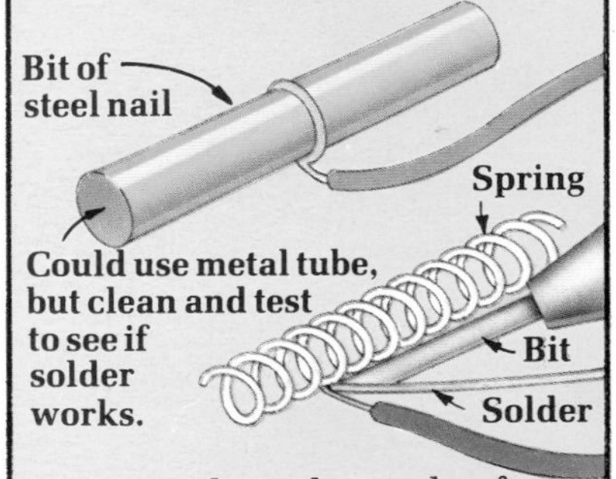

Strip and tin the ends of two 400mm wires. Cut a 50mm length of steel nail, clean it with sandpaper (or it will not solder) and solder a wire to it. Twist the other wire round a spring and solder.

2

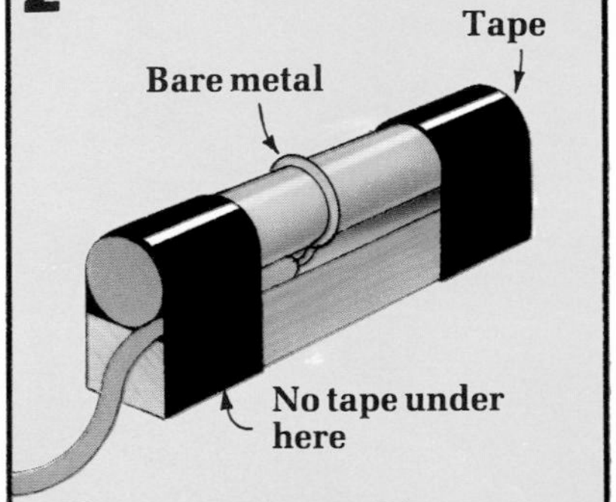

Cut a 6mm spar 50mm long. Tape the nail on top of the spar with the wire running under the tape as shown. Make sure the centre part of the nail is bare and that there is no tape under the balsa.

3

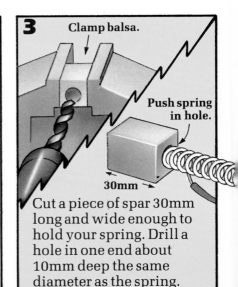

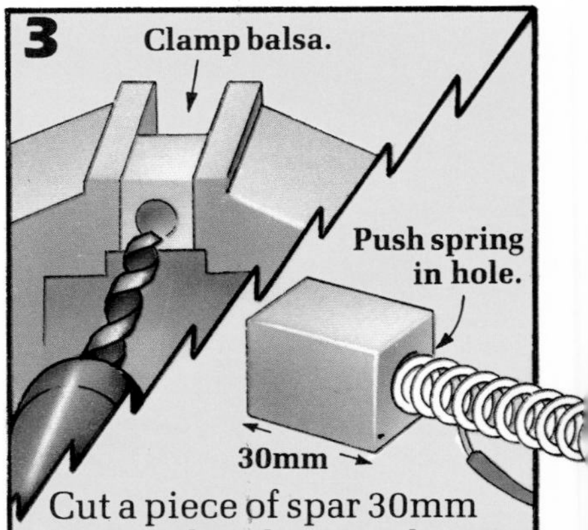

Cut a piece of spar 30mm long and wide enough to hold your spring. Drill a hole in one end about 10mm deep the same diameter as the spring. Then push the spring in the hole so it sticks out.

*See page 22 to find out how to solder.

Gripper closed sensor

You may find it easier to remove gripper.

1

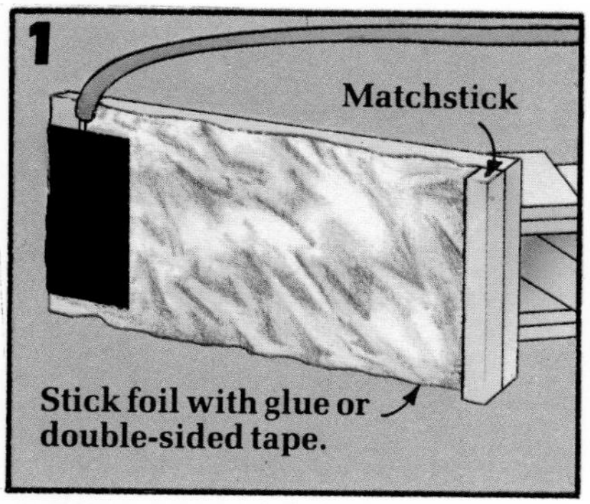

Stick tin foil to the face of the right-hand jaw. Strip a 600mm wire and tape one end to the foil. Glue a piece of matchstick to the jaw as shown.

2

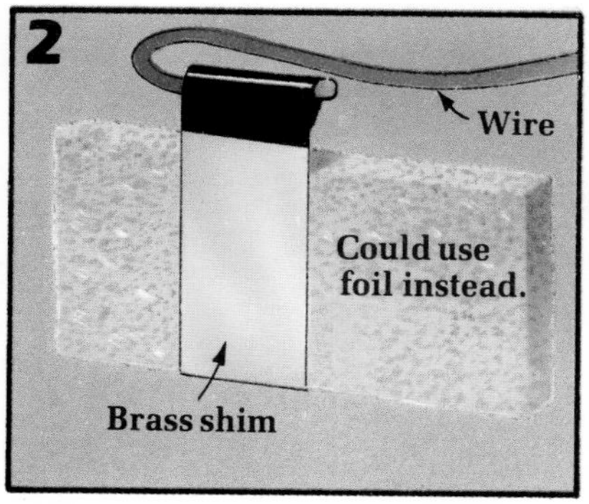

Cut a piece of foam 55mm × 30mm and a strip of brass shim 15mm × 40mm. Glue or tape the shim to the foam. Strip a 600mm wire and tape it to the shim.

3

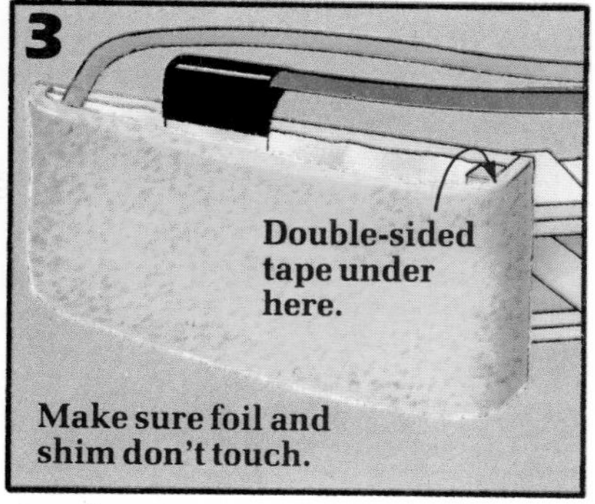

Stick a thin strip of double-sided tape to either end of the jaw. Stretch the foam and stick it to the end of the jaw and the matchstick.

4 Gently bend the wires from the jaw and tape them to the gripper.

5 Glue a piece of foam to the other jaw to help the gripper hold things.

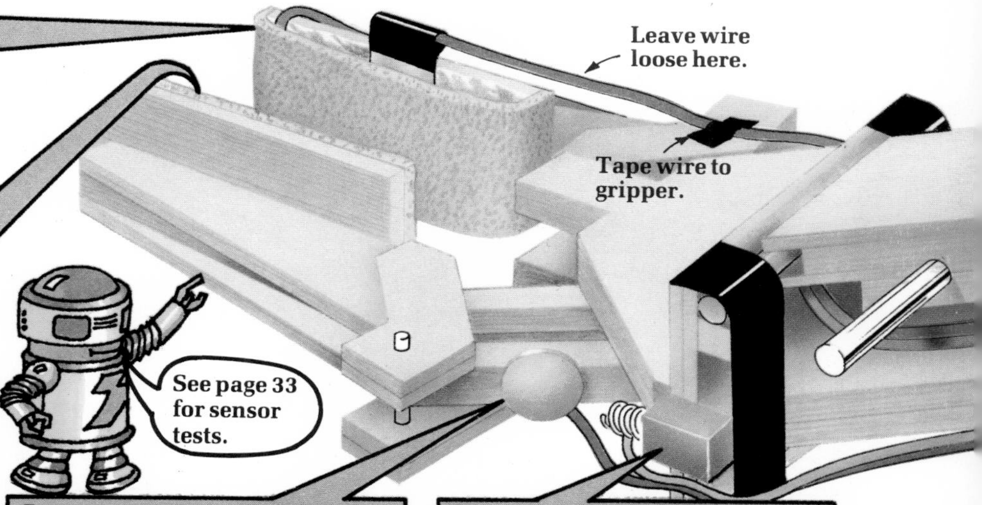

4

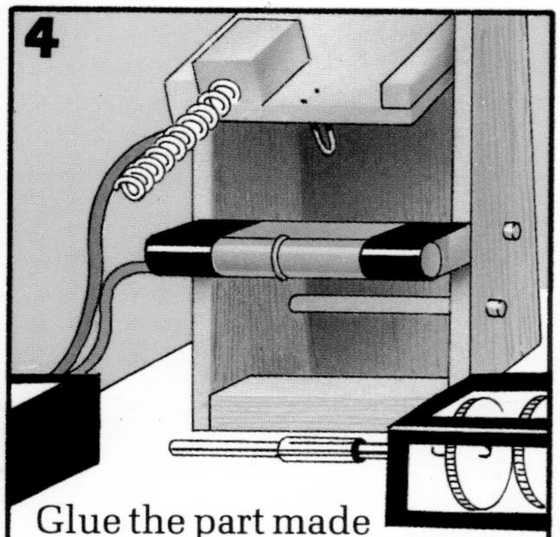

Glue the part made in step 2 to the back of the shoulder, about 25mm up from the base. Then glue the part made in step 3 to the lower part of the arm so the spring touches the nail when the arm is fully up.

1 Gripper open sensor

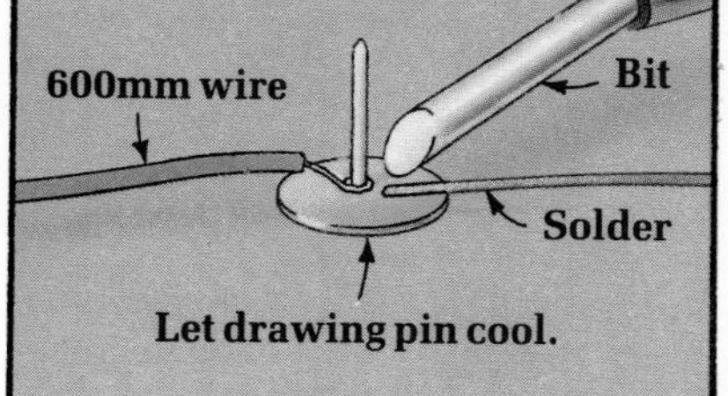

Strip and tin a 600mm wire and solder one end to a drawing pin. Push the pin in the gripper to line up with step 2.

2 Follow step 3 in the yellow boxes to makes this part.

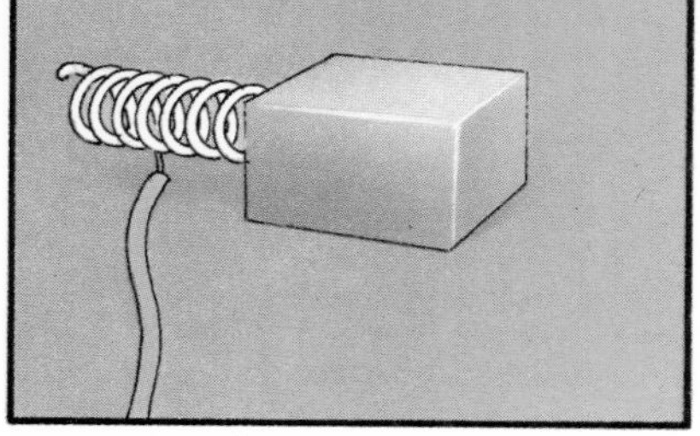

Make another spring part. Glue it to the side of the wrist so it touches the drawing pin when the gripper is fully open.

How to make a light sensor

You can find out here how to make a light sensor, as an added extra for any version of the robot.* This enables you to instruct the robot to follow a line drawn on a big sheet of paper, or "look" for bright objects. There is an example of how to instruct the robot and light sensor on page 39. You can also find out how to make an optional on/off light on the opposite page.

Materials you need

Sensor templates A to F, 1½mm balsa sheet, Velcro, sticky tape, dark paper, 2m bell wire, LDR, LED, 10 hole connector block, 330Ω resistor, 2 small self-tapping or wood screws, soldering and tool kits.

How light sensor works

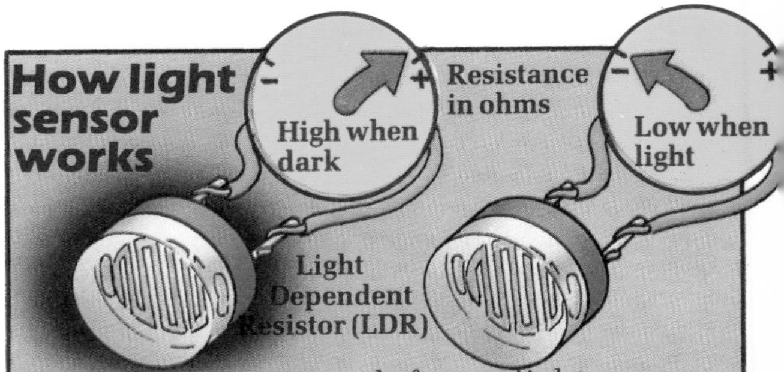

The light sensor is made from a light dependent resistor, connected to the computer via a circuit, where there is a knob to adjust its sensitivity to light. The LDR sends an "on" message to the computer when light shines on it and an "off" message when it is dark.

Making the light sensor

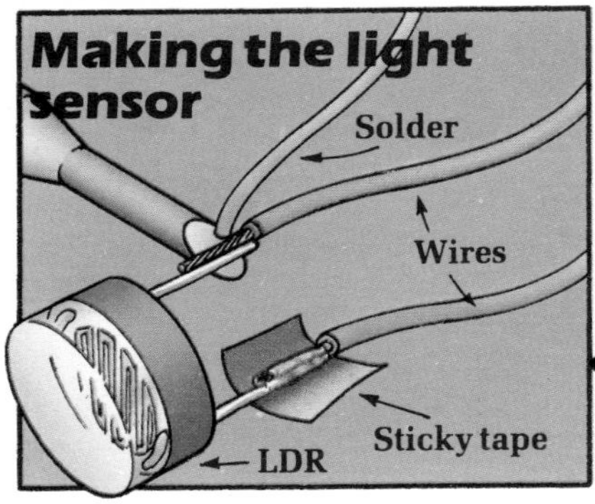

Strip and tin each end of two 500mm wires. Then solder one end of each wire to the legs of an LDR.

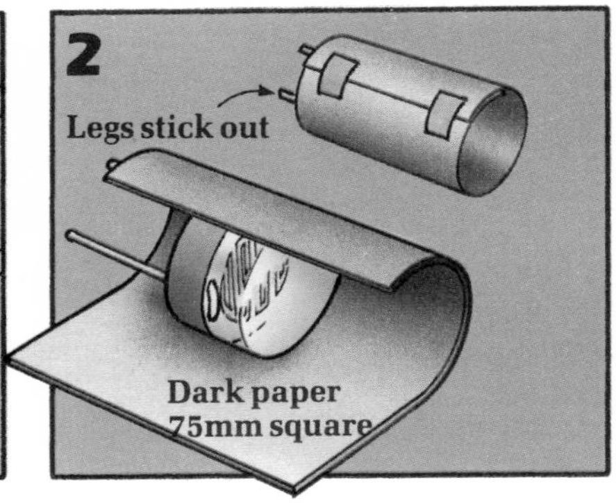

Cut out a piece of dark coloured paper or thin card and make a tube round the LDR. Stick the tube with tape.

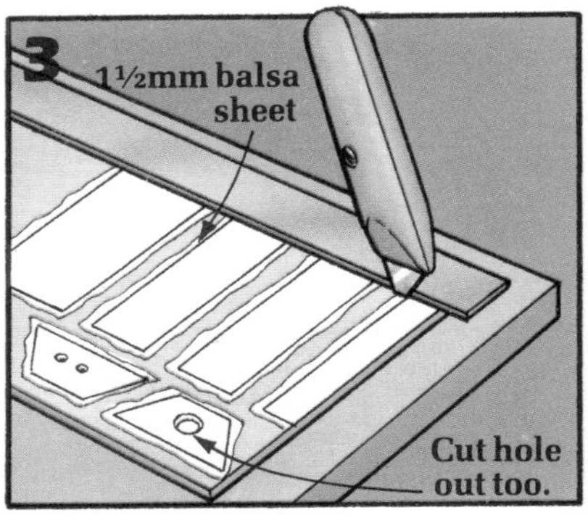

Cut out light sensor templates A to F and glue them to 1½mm thick balsa sheet. Cut each piece out with a craft knife.

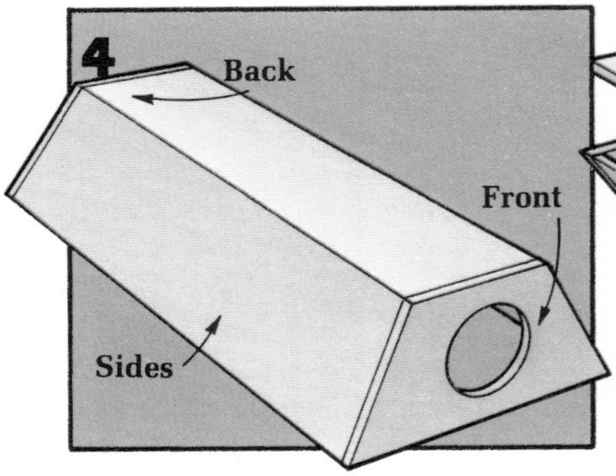

Glue the front, back, top and sides together as shown. Hold each part while the glue sets. Don't glue the bottom on yet.

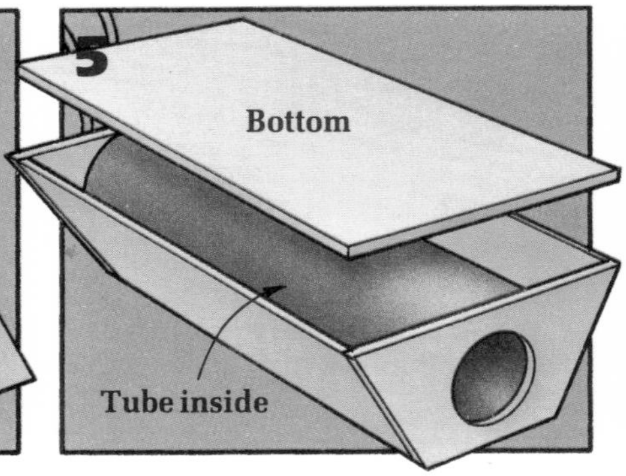

Place the paper tube inside the box with the wires poking out of the holes at the back. Then glue the bottom on.

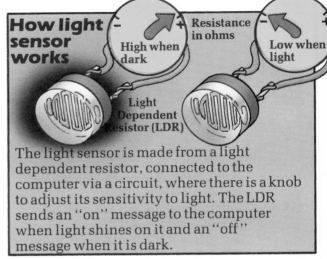

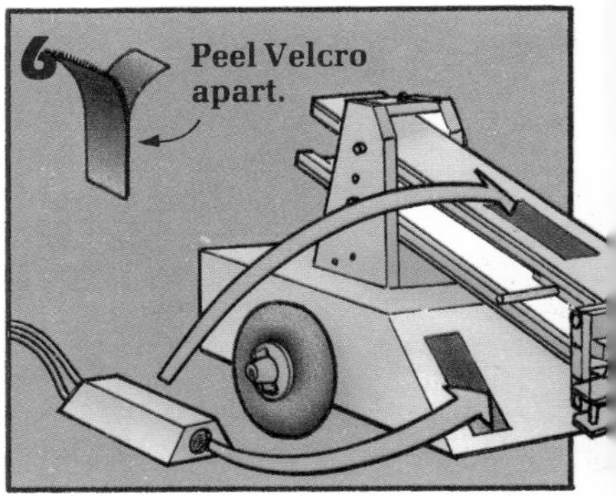

Cut two bits of Velcro 70mm long. Separate the Velcro, glue one bit under the sensor and two bits to the robot.**

*If you don't make the light sensor, leave out the potentiometer in the circuit on page 32.

**See opposite to find out how to connect wires up.

Connecting up the robot's wires

All the wires from the robot's sensors, motors and LED (if you add one) are connected into a connector block. This makes the robot ready to connect to the switching circuit explained on page 30.

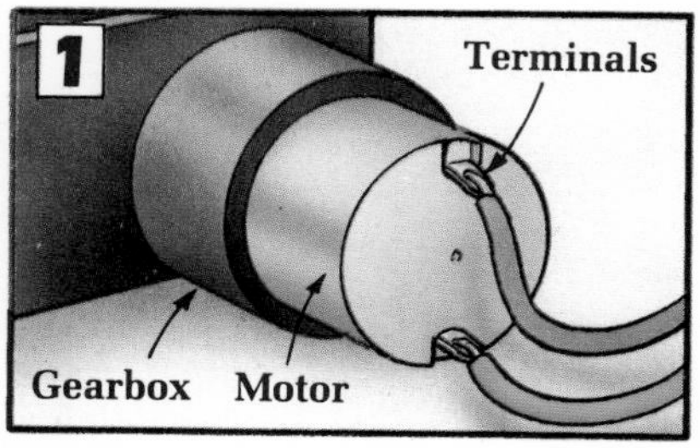

Strip and tin one end of eight 300mm long wires. Solder the tinned end of each wire to each of the motor terminals.

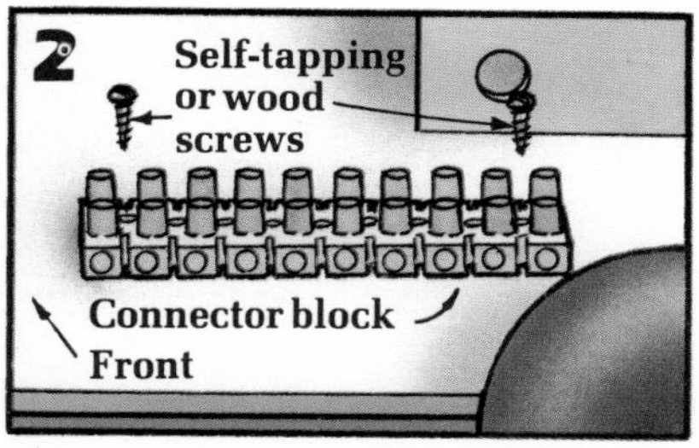

Screw a connector block with ten pairs of holes to the base.

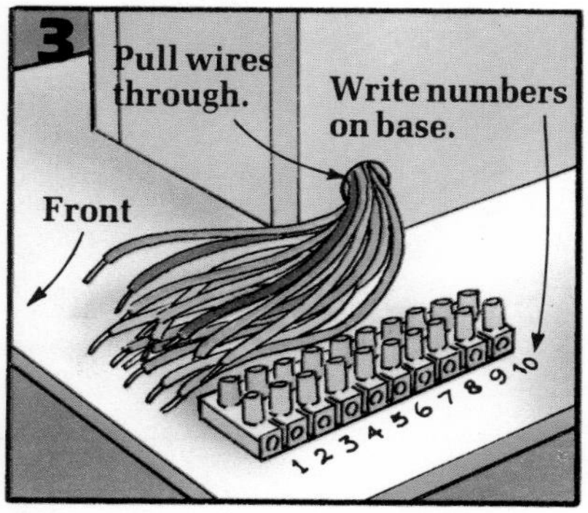

Write numbers by the side of the connector block. Thread any loose wires through the hole in the shoulder.

On/off light

1 Strip and tin the ends of two 250mm long wires and one 100mm wire. Carefully twist one end of each 250mm wire round the legs of an LED and solder them in place. Then wrap sticky tape round each join. Note which wire is soldered to the positive leg.

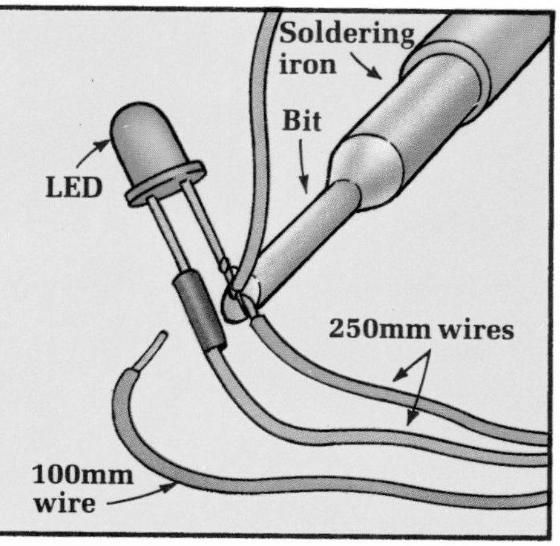

2 Feed the LED wires through the hole in the top of the shoulder and push the LED in as shown. Fix the LED in with the plastic sleeve supplied with it (not all have one) or with plasticine.

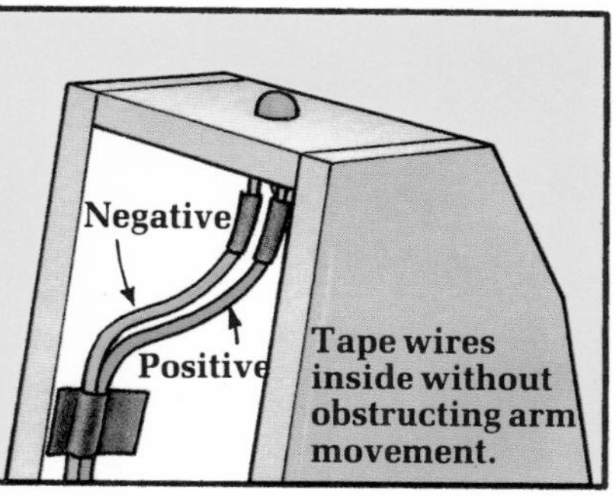

3 Feed the wires through the bottom of the shoulder. Tape a 330Ω resistor where marked on the base template. Then solder the wire from the positive leg of the LED to one end and the 100mm wire to the other.

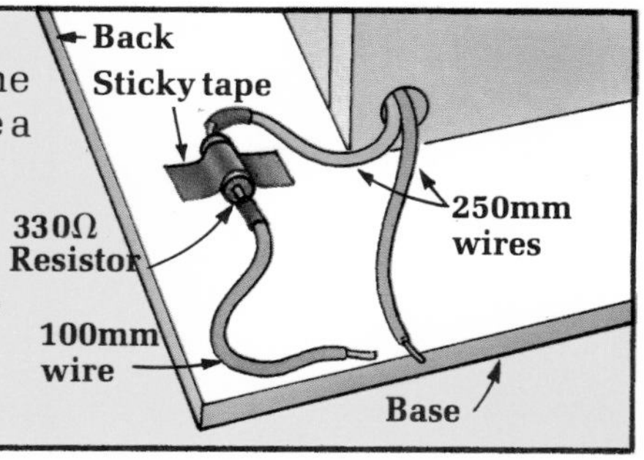

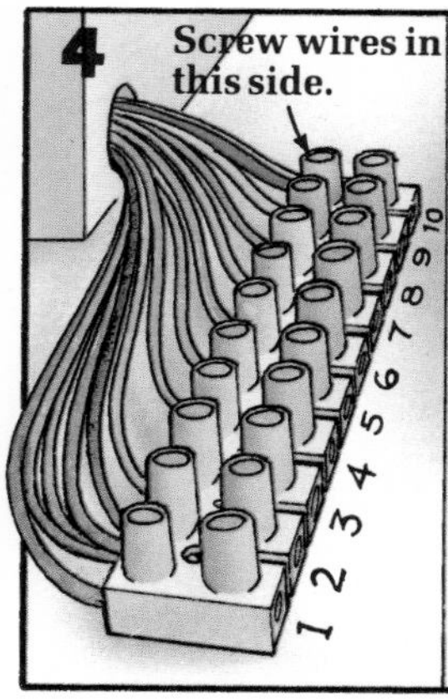

1	Wire from LED positive leg via 330Ω resistor.
2	Wire from LED negative leg and one from each sensor (including light).
3	One wire from each arm sensor.
4	One wire from each gripper sensor.
5	One wire from light sensor.
6	One wire from arm motor.
7	One wire from gripper motor.
8	One wire from left drive motor.
9	One wire from right drive motor.
10	One wire from each motor.

Screw the wires into the connector block as shown in this chart. There are 18 wires with the LED, 16 without.

Use the same hole numbers as shown above for all versions of the robot. Do this carefully, checking each wire as you go.

Making a switching circuit

These instructions show how to make the switching circuit which connects the robot to the computer. It is important to follow them very carefully, as one mistake could damage the circuit or your computer. The template on page 41 helps identify where components go.

Things you need

Relays, transistors, diodes, resistors, potentiometer (see page 40 for exact types), bell wire or ribbon cable, Veroboard, soldering kit and tool kit, including drill bit (about 5mm).

Important

See page 41 to check relay pin numbers. DO NOT BEGIN UNTIL YOU HAVE DONE THIS.

1

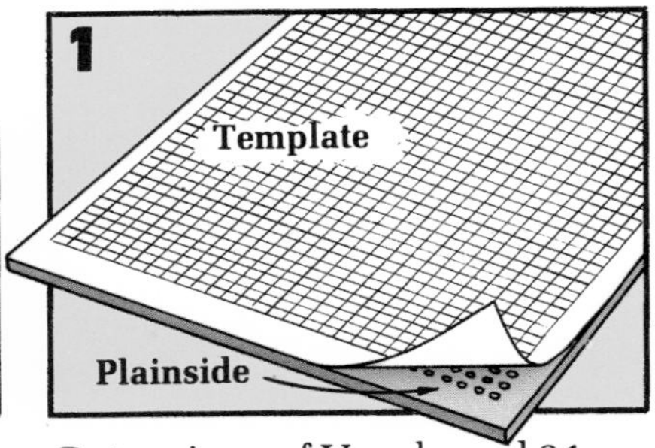

Cut a piece of Veroboard 31 tracks by 50 holes long. Photocopy or accurately trace the circuit template on page 41. Cut it out with scissors and glue it to the plainside of the Veroboard so the crosses line up with the holes.

2

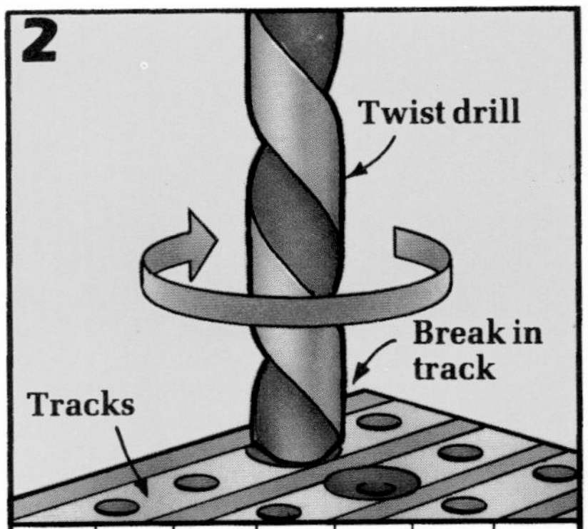

S5	R6	Z6	a6	T7	Z13	a13
b13	c13	Q16	Z16	a16	Z23	a23
b23	c23	N24	P26	Z26	a26	C31
D31	E31	F31	G31	H31	I31	J31
Z33	a33	b33	c33	G36	H36	O36
Z36	a36	N40	O40	P40	Q40	R40
S40	I47	O47	P47	Q47	R47	

Turn the Veroboard over and cut the track at the holes shown in the chart. Identify the holes by pushing a pencil point through the template. Clamp the board and cut the track with a drill bit. Check the track is cut through.

3

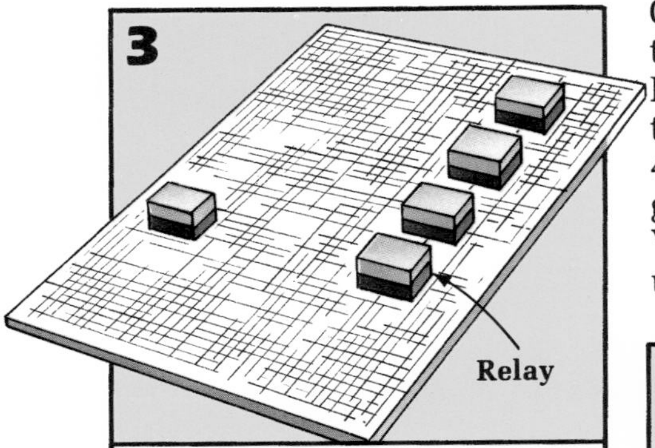

Relay 1						
Pin No.	1	2	3	4	5	6
Hole	a8	Z8	V8	a5	Z5	V5
Relay 2						
Pin No.	1	2	3	4	5	6
Hole	a18	Z18	V18	a15	Z15	V15
Relay 3						
Pin No.	1	2	3	4	5	6
Hole	a28	Z28	V28	a25	Z25	V25
Relay 4						
Pin No.	1	2	3	4	5	6
Hole	a38	Z38	V38	a35	Z35	V35
Relay 5						
Pin No.	1	2	3	4	5	6
Hole	H38	G38	C38	H35	G35	C35

Push the legs of five relays through the holes shown in the chart. Bend the legs out slightly as you go, then solder them carefully to the track.

4

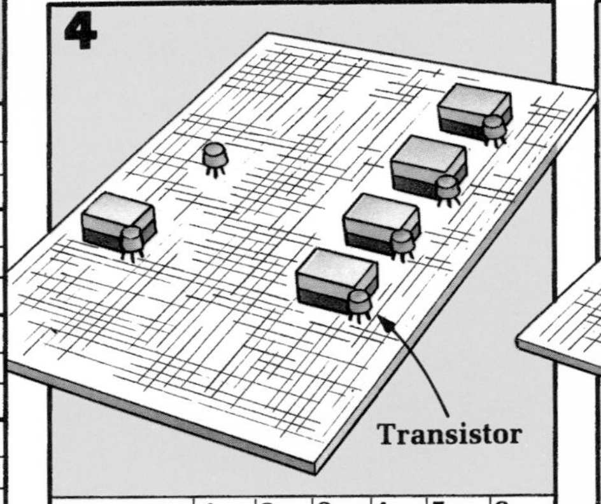

	1	2	3	4	5	6
Collector	a10	a20	a30	a40	H40	G23
Emitter	c10	c20	c30	c40	J40	I23
Base	b9	b19	b29	b39	I39	H22

Push the legs of six transistors through the holes shown in the chart. Solder transistors to the track quickly to avoid heating them.

5

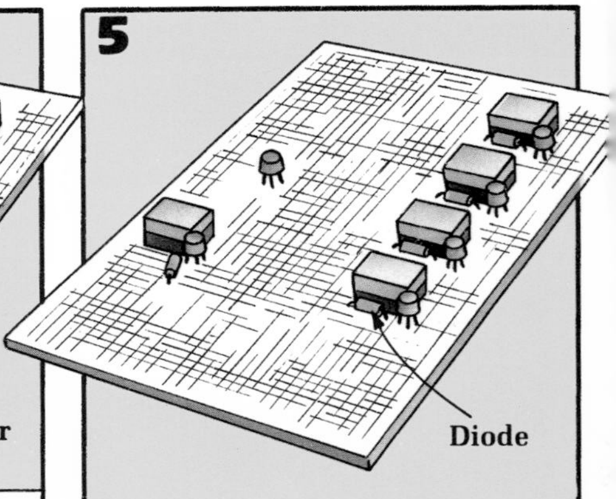

T9	a9	T19	a19	T29	a29
T39	a39	G40	H45		

Push the legs of five diodes through the holes shown in the chart, with the striped end in the hole marked in yellow. Then solder to the track.

Transistors: Ask your supplier which leg is nearest spot or other mark on case (see page 40).

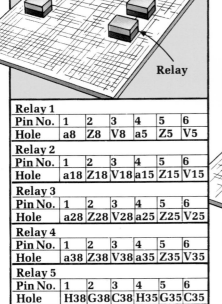

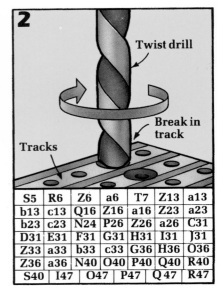

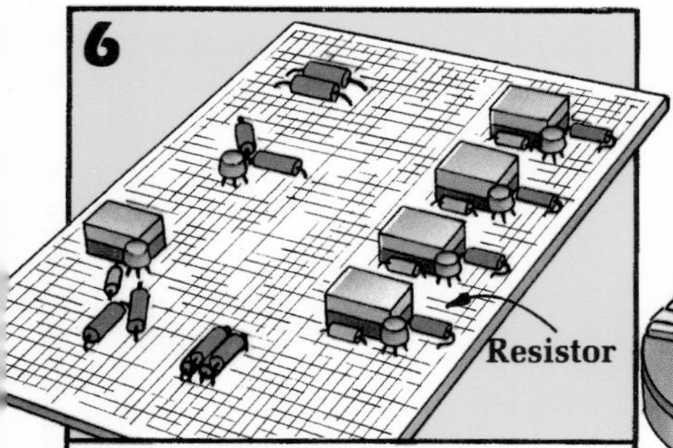

1KΩ Brown/Black/Red			
E17/G22	H20/N20		
2K2Ω Red/Red/Red			
E6/M6	E8/L8	I43/I48	O43/O48
P43/P48	Q43/Q48	R43/R48	
6K8Ω Blue/Grey/Red			
b8/e8	b18/e18	b28/e28	
b38/e38	I41/K46		

Push the legs of fourteen resistors through the holes shown in the chart, bend the legs out and solder them to the track.

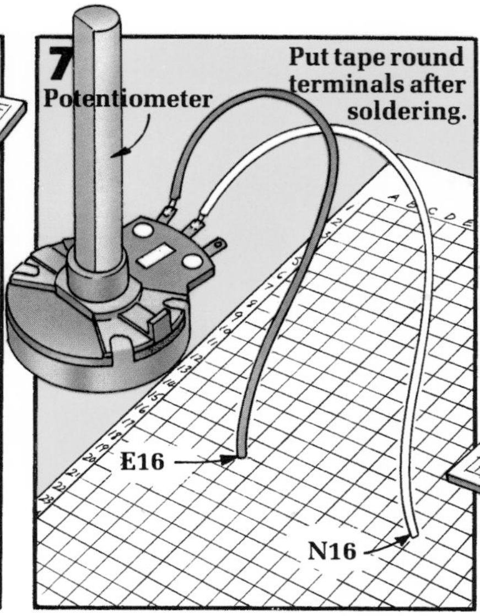

Strip and tin the ends of two 150mm wires. Solder them to the centre and outside tag of a 100K potentiometer. Solder one wire in E16 and one in N16.

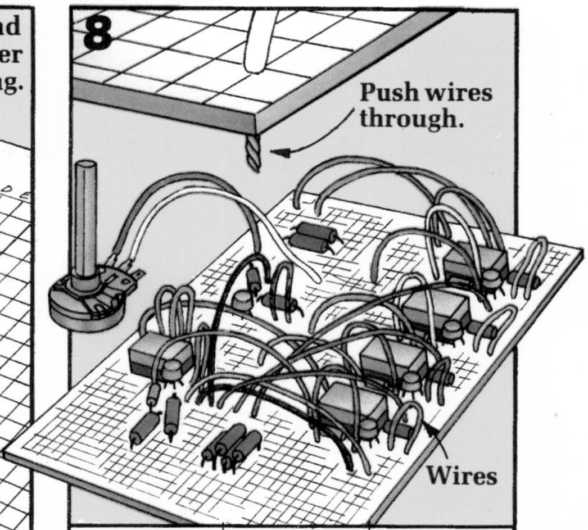

c11/e11	c21/e21	c31/e31
c41/e41	J39/K39	T10/Z10
T20/Z20	T30/Z30	T40/Z40
G39/T38	R4/Z4	Q14/Z14
P24/Z24	O34/Z34	O41/b44
P41/b32	Q41/b22	R41/b12
A3/S3	E4/X4	I24/K24
G25/N41	K42/e42	C34/V34
D34/G34	E34/H34	

Strip and tin the ends of twenty-six 150mm wires. Solder them as loops between the holes shown in the chart. Check them as you go.

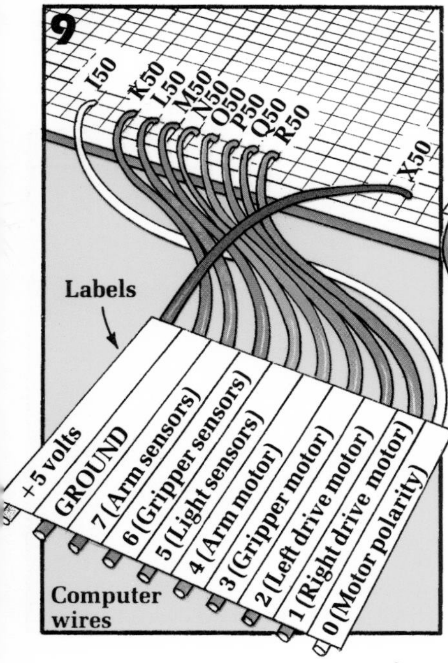

Cut 500mm of 10-way ribbon cable or ten lengths of bell wire. Strip and tin the ends. Tape labels to them and solder them in the holes shown.

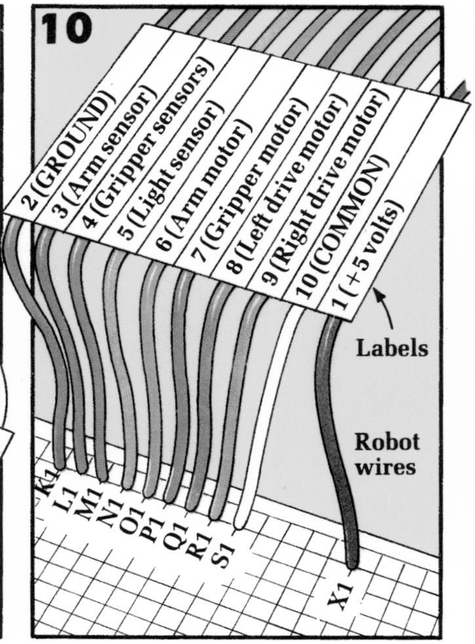

Cut 2m of 10-way ribbon cable or ten lengths of bell wire. Strip and tin the ends. Tape labels to them and solder them in the holes shown.

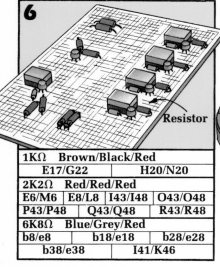

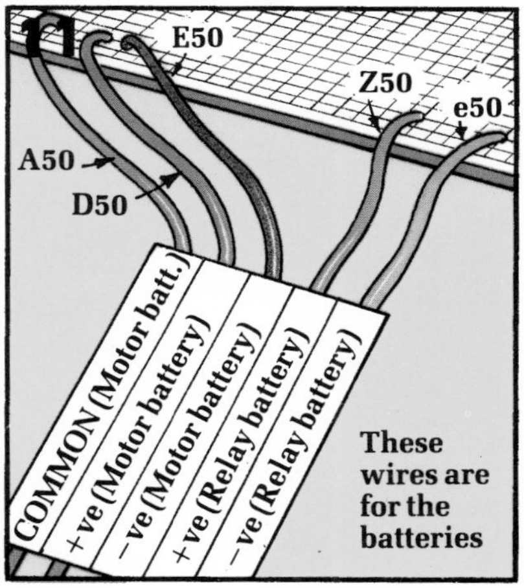

Strip and tin the ends of five 200mm wires. Tape labels to them and solder one end of each wire into the holes shown above.

IMPORTANT: Check no tracks are joined, or you could damage your computer.

Connecting the robot, circuit and computer

These pages explain how to connect the switching circuit to the robot, batteries and computer. It is very important to read the instructions before connecting up, and then follow them carefully, or you may damage your computer.

Relay power

Instead of a 12 volt battery for the relay power, you could join two 6 volt batteries together as shown above.

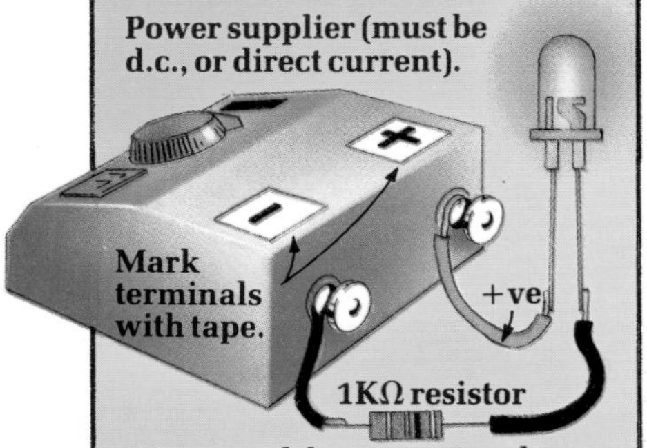

You could use a 12 volt power supplier from a slot car set. Not all power supplier terminals are marked +ve (positive) and −ve (negative). Test which is which with a 1K resistor and an LED as shown. The LED only lights up when its negative leg is connected to the negative terminal.

1

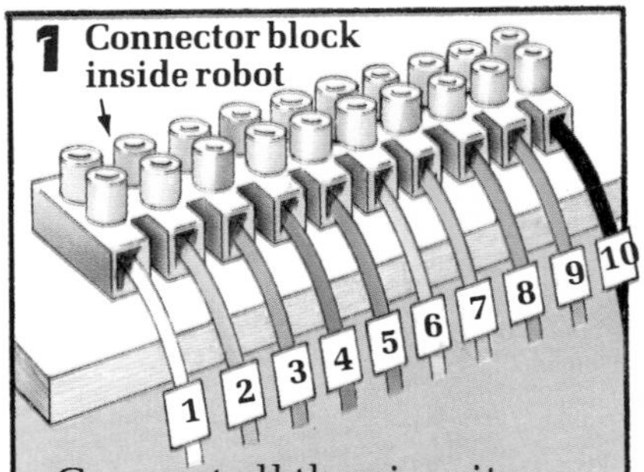

Connect all the circuit robot wires into the connector block fixed to the robot as shown above. Check that the wires are in the correct holes.

2

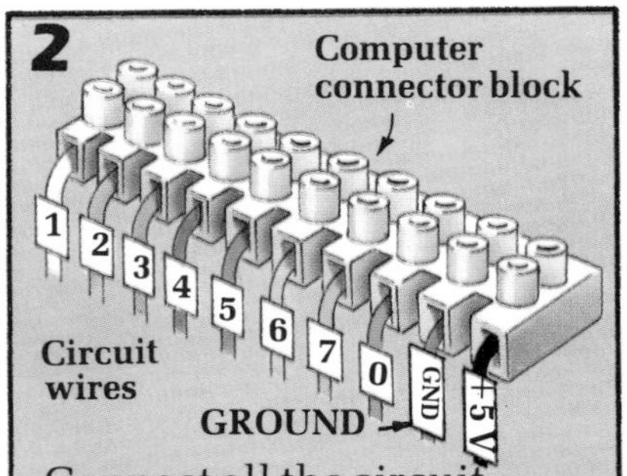

Connect all the circuit computer wires into another connector block with ten pairs of holes as shown above. Label the block with tape as shown.

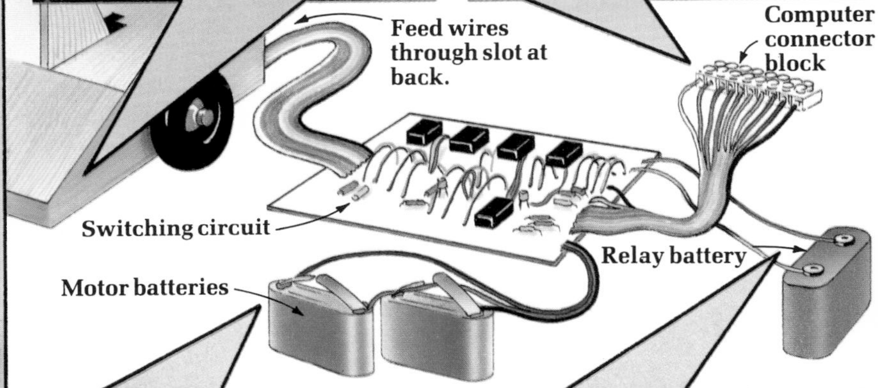

3

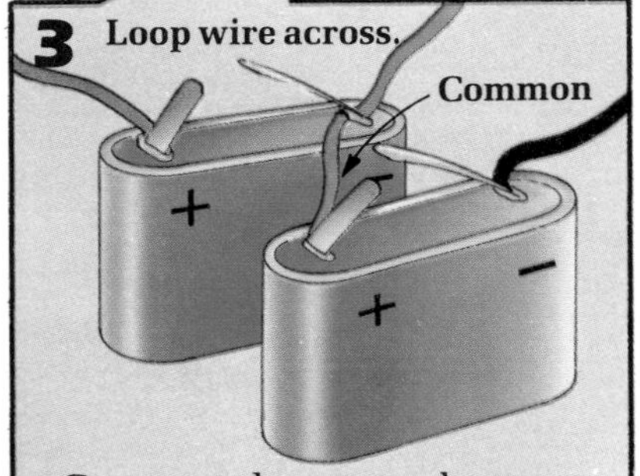

Connect the motor battery wires to two batteries as shown. Connect the wire labelled COMMON between the +ve and −ve battery terminals, using an extra wire. Use two batteries of the same voltage as your motors (e.g. 3 volt motors need two 3 volt batteries).

4

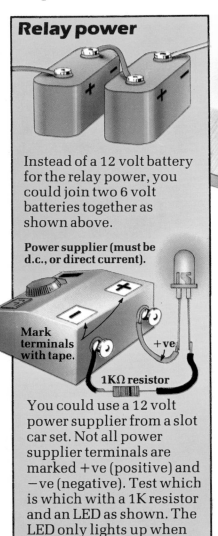

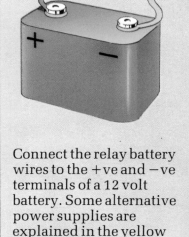

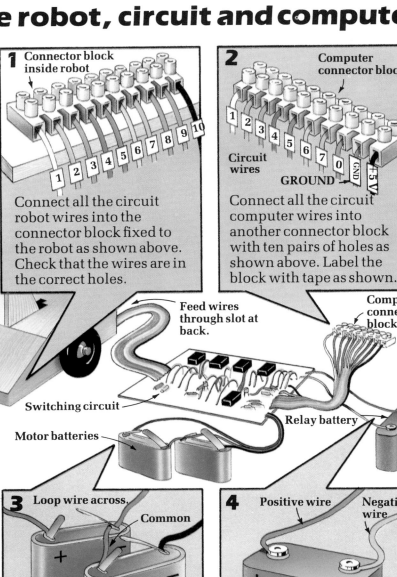

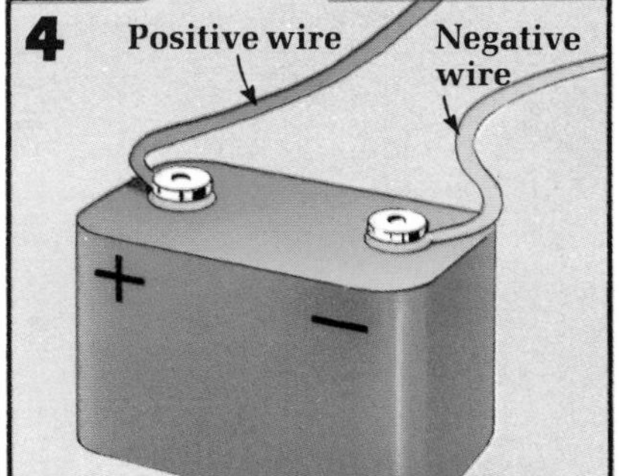

Connect the relay battery wires to the +ve and −ve terminals of a 12 volt battery. Some alternative power supplies are explained in the yellow panel on the left.

Turn the page before switching your computer on.

Test bench report

Circuit

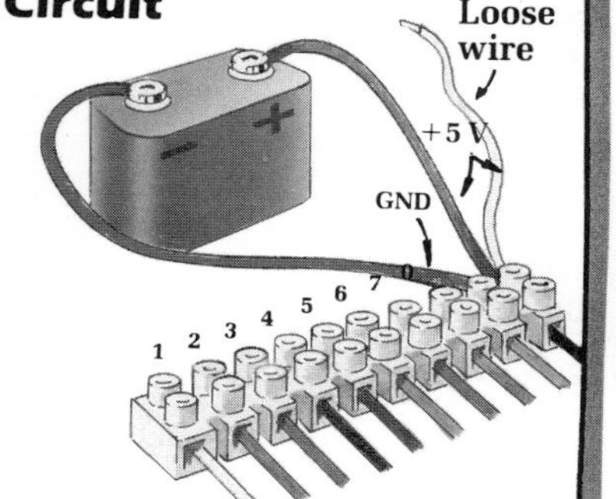

Test the circuit before connecting to your computer to see if it switches the motors on and off. Strip three wires and connect them as shown to the computer connector block and a 6 volt battery (the other batteries must be connected too). Touch the loose wire in holes 1,2,3 and 4, one at a time. You should hear a relay click and one motor start as you try each one. Remove the loose wire.

Sensor test

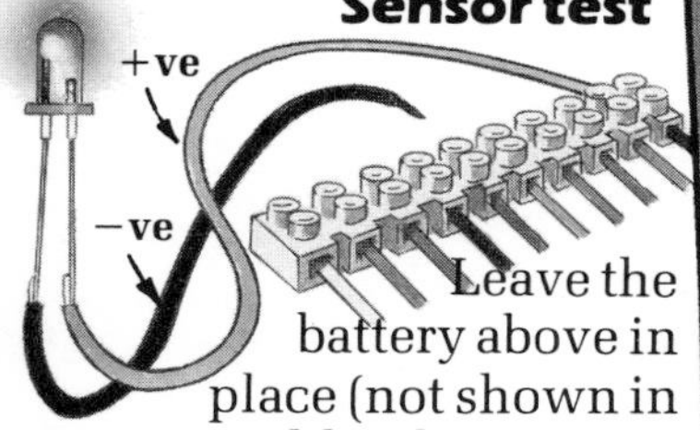

Leave the battery above in place (not shown in picture). Solder three wires and a 330Ω resistor to an LED. Put the +ve wire in the 5 volt hole. Touch the other wire in these holes, while manually moving the robot (remove wires and battery afterwards):

Hole	What should happen
3	**LED on when arm fully up or down.**
4	**LED on when gripper fully open or closed**
5	**LED off when light sensor pointed at bright light.**

BBC Model B

User port pin number	Label on computer connector block
1	+5 volts
19	GROUND
6	0 (motor polarity)
8	1 (right drive motor)
10	2 (left drive motor)
12	3 (gripper motor)
14	4 (arm motor)
16	5 (light sensor)
18	6 (gripper sensor)
20	7 (arm sensor)

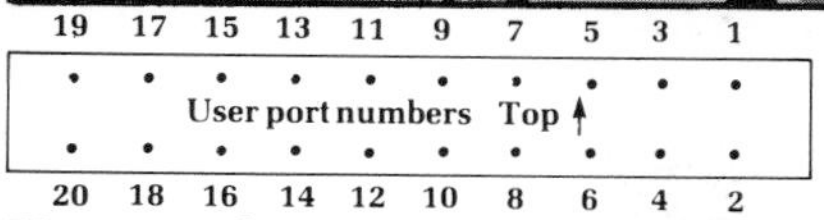

You need a special plug called a 20-pin I.D.C. plug to connect the circuit to the socket marked "User port" on the BBC. Buy one with a length of ribbon cable attached, and connect to the computer connector block as shown in this chart.

Commodore 64 and VIC 20

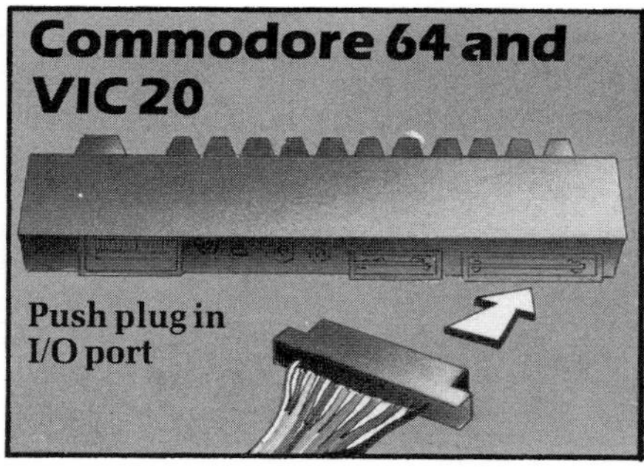

User I/O pin numbers	Label on computer connector block
2	+5 Volts
N	GROUND
C	0 (motor polarity)
D	1 (right drive motor)
E	2 (left drive motor)
F	3 (gripper motor)
H	4 (arm motor)
J	5 (light sensor)
K	6 (gripper sensor)
L	7 (arm sensor)

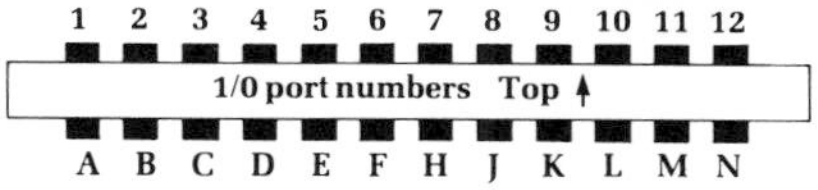

Use a plug called an edge connector for these computers. Cut a short length of ribbon cable and strip and tin the wires at both ends. Connect the wires to the edge connector and computer connector block as shown above.

ZX Spectrum

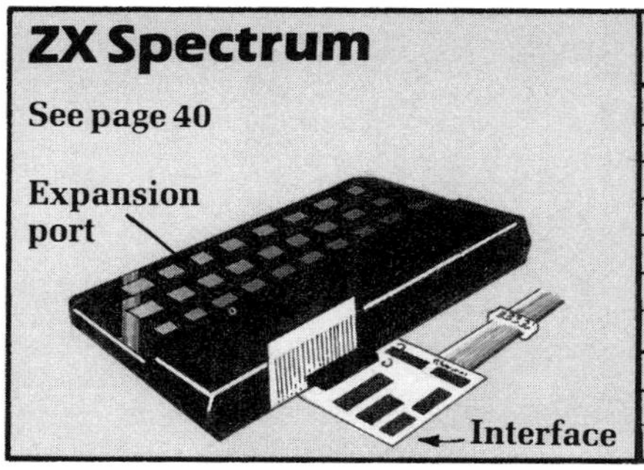

Interface pin number		Label on computer connector block
+5 V		+5 Volts
−0 V		GROUND
0	Output lines	0 (motor polarity)
1	Output lines	1 (right drive motor)
2	Output lines	2 (left drive motor)
3	Output lines	3 (gripper motor)
4	Output lines	4 (arm motor)
5	Input lines	5 (light sensor)
6	Input lines	6 (gripper sensor)
7	Input lines	7 (arm sensor)

You need a special interface circuit which plugs into the expansion port. There are a number of types and they give eight input and eight output lines. Following the interface instructions, connecting a short length of ribbon cable between the interface and computer connector block as shown above.

Do these checks if the tests don't work:

- Check all wires are in the correct holes.
- Check soldering, re-solder any "dull" looking joints.
- Check track breaks against the chart on page 30.
- Try another LED, making sure you identify the legs correctly.
- Are your batteries fresh?
- **If everything fails, see page 48.**

Test programs

Here are some tests to check the robot is working correctly before typing in the main program over the page. There is also a "start-up" routine to do each time you use the robot. This makes sure the motors do not start before you are ready.

"Start-up" routine

1 Connect the robot and computer as shown on page 32. Disconnect the battery COMMON wire. Switch the computer on, then type in the lines on the right.

2 Re-connect the COMMON wire.

3 Press RETURN (ENTER on Spectrum). Now you are ready to type in the main program or test programs.

***BBC**

```
?&FE62=31
?&FE60=0
```

***VIC 20**

```
POKE 37138,31
POKE 37136,0
```

***C64**

```
POKE 56579,31
POKE 56577,0
```

***Spectrum**

```
either POKE port number,0  or  OUT
port number,0
```

The commands you use for the Spectrum depend on the interface you buy. See page 40.

Motor Test

1 Type in the program below for your computer. Type RUN when you are ready.

***BBC**

```
10 ?&FE60=0
20 PRINT "TRY A NUMBER"
30 INPUT X:?&FE60=X
40 I$=GET$:GOTO 10
```

***VIC 20**

```
10 POKE 37136,0
20 PRINT "TRY A NUMBER"
30 INPUT X:POKE 37136,X
40 GET A$:IF A$="" THEN GOTO 40
50 GOTO 10
```

***C64**

C64 users use VIC-20 version but replace 37136 with 56577

***Spectrum**

```
10 either POKE port number,0 or OUT port number,0
20 PRINT "TRY A NUMBER":INPUT X
30 either POKE port number,X or OUT port number,X
40 PAUSE 50
50 IF INKEY$="" THEN GOTO 50
60 GOTO 10
```

Port number depends on interface.

2 The computer makes the motors run in one direction or another, depending upon which way round the motor terminal wires are connected. When "TRY A NUMBER" appears on the screen, type the numbers shown in the chart, one at a time. Watch each motor to check it runs in the direction shown. Press RETURN (ENTER on Spectrum) to stop the motor. Swap over the terminal wires of any motor running the wrong way.

Number to type in program	Motor to watch	Direction robot should go
2	Right drive	Forwards
3	Right drive	Backwards
4	Left drive	Forwards
5	Left drive	Backwards
8	Gripper	Close
9	Gripper	Open
16	Arm	Up
17	Arm	Down
0	To stop the test	

Sensor test

This checks to see if the sensors are working properly. Check all the wires on the sensors if any part of the test fails.

1 Unhook the fishing line and elastic bands on the arm and gripper so you can move them by hand. Then type the program for your computer and type RUN when you are ready.

***BBC**

```
10 PRINT ?&FE60
20 GOTO 10
```

***VIC 20**

```
10 PRINT PEEK(37136)
20 GOTO 10
```

***C64**

```
10 PRINT PEEK(56577)
20 GOTO 10
```

***Spectrum**

```
10 either PRINT PEEK(port number)
   or PRINT IN (port number)
20 GOTO 10
```

Use these programs for light sensor test too.

Arm/gripper position	Number on screen
Arm half up	A number
Arm up	Number above –128
Arm down	
Gripper half open	A number
Gripper closed	Number above –64
Gripper open	

2 Hold the arm and gripper in each of the positions shown in the chart above. If the sensors are working correctly the numbers in the chart will appear on the screen for each position.

***To stop the program type: BBC: ESCAPE then NEW C64/VIC20: RUN/STOP then NEW SPECTRUM: BREAK then NEW**

Light sensor

1 Type in the sensor test program. Point the light sensor at a diffused source of light, towards the window, for example.

2 Twist the shaft of the potentiometer connected to the switching circuit. This adjusts the sensitivity of the light sensor. Make adjustments with the sensor pointing towards the light, until you see a number on the screen. When you put your hand over the sensor, you should see the same number minus 32. Adjust the light sensor according to what you use it for. For example, to get the robot to follow a black line drawn on paper, position the sensor on the front of the robot. Then move the robot from side to side over the line, adjusting the potentiometer at the same time to read a low number while pointing at the line, and a higher number while pointing at the white paper.

Computer control

The computer program over the page controls the motors according to instructions you give the computer. It does this by sending and receiving electrical voltages (about 5 volts) through the wires connected between the switching circuit and the computer's port (usually called a parallel input/output port). The port has eight separate wires inside, called lines.

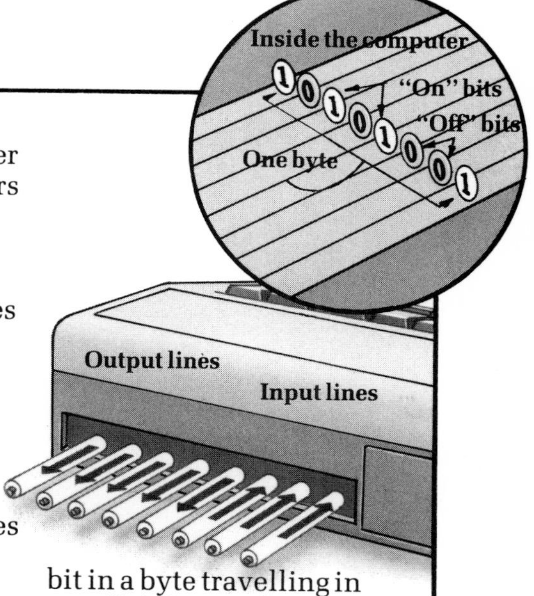

Inside the computer, each instruction is converted into a number in binary code. Binary is a number system using only two digits, 0 and 1, to represent any decimal number. Individual digits are called bits (short for binary digits), and they are represented by an electrical pulse (about 5 volts) for a 1 and no pulse for 0. Most home computers use groups of eight bits, called bytes, to represent each part of an instruction. The parallel input/output port is like an eight lane road, with each bit in a byte travelling in parallel along the separate lines.

Some lines, called output lines, send messages to the motors, while others, called input lines, receive messages from the sensors. The program sets five lines to output and three to input. One output line controls the direction the motors turn, or their polarity, and each of the other four switch one motor on or off. The gripper, arm and light sensors send messages to the computer separately along the three input lines.

Program changes

The main program works for the BBC computer. These are the changes you need to make to the main program over the page for the VIC 20, C64 and Spectrum computers.

■ C64 ▲ VIC 20

```
■       10 POKE 56579,31:OP=56577
▲       10 POKE 37138,31:OP=37136
■ ▲     30,230,330,990 Replace CLS with PRINT CHR$(147)
■ ▲ 800 GET I$:IF I$<>"" THEN GOSUB 830:PF=1:RETURN
■ ▲ 830 POKE OP,0:RETURN
■ ▲ 840 POKE OP,G:RETURN
■ ▲ 860 IF (PEEK(OP) AND TN)=0 THEN SF=1
■ ▲ 900 FOR T=1 TO 200:NEXT T
```

Spectrum

```
 10 LET OP=the number of your port
100 GOSUB 330*(A=1)+230*(A=2)+120*(A=3)
170 GOSUB 480*(Q<8)+500*(Q=8)+540*(Q=9)+620*(Q=10)+640*(Q=11)
190 GOSUB 680*(Q=12)+690*(Q=13)+710*(Q=14)+730*(Q=15)+750*(Q=16)
270 PRINT "  ";I;".  ";:IF M(I)>0 THEN PRINT V$(M(I));
375 IF LEN(C$)<6 THEN LET C$=C$+" ":GOTO 375
800 IF INKEY$<>"" THEN GOSUB 830:LET PF=1:RETURN
830 either POKE OP,0 or OUT OP,0:RETURN
840 either POKE OP,G or OUT OP,G:RETURN
850 LET SF=1
860 LET D= either PEEK OP or IN OP
862 FOR I=7 TO 5 STEP -1:LET Z=2^I
864 IF D>=Z THEN LET D=D-Z:IF Z=TN THEN LET SF=0
866 NEXT I
900 FOR T=1 TO 100:NEXT T
920 DIM M(40):DIM D(40):DIM P(11):DIM V$(17,6)
```

Robot control program

This program is written in BASIC. No changes are needed for the BBC computer. Changes for other computers are on page 35.

"Start-up"
```
10 ?&FE62=31:OP=&FE60
```
— Sets data direction register
```
20 GOSUB 830:GOSUB 920:GOSUB 990
```
— Sets robot to starting position and reads data.

Menu
```
30 CLS
40 PRINT "ROBOTROL MENU":PRINT
50 PRINT "DO YOU WANT TO "
60 PRINT "1. ENTER STEPS"
70 PRINT "2. DISPLAY":PRINT "3. GO"
80 PRINT "TYPE A NUMBER"
```
— Prints menu
```
90 INPUT A:IF A<1 OR A>3 THEN GOTO 90
```
— Gets user's choice (A).
```
100 ON A GOSUB 330,230,120
```
— Branches to routine depending on choice.
```
110 GOTO 30
```

Run steps
```
120 LET PC=1:LET PF=0
```
— Sets PC (step number) to 1. Sets PF (panic button flag) to 0.
```
130 LET Q=M(PC)
```
— Sets Q to number of command stored in V$ as recorded in M for this step.
```
140 IF Q=17 THEN GOSUB 830:RETURN
```
— Stops motors and returns to menu if command is HALT (command 17 in array V)
```
150 IF Q=0 THEN GOTO 210
160 IF Q>11 THEN GOTO 190
170 ON Q GOSUB 480,480,480,480,480,480,480,500,540,620,640
```
— Branches to relevant section depending on command.
```
180 GOTO 200
190 ON Q-11 GOSUB 680,690,710,730,750
200 IF PF=1 THEN RETURN
```
— Returns to menu if PF (panic button flag) =
```
210 LET PC=PC+1:IF PC>40 THEN GOSUB 830:RETURN
```
— Increases step number (PC) for next step and goes back.
```
220 GOTO 130
```

List steps
```
230 CLS
240 PRINT "FROM WHICH STEP":INPUT S
250 LET EL=S+15:IF EL>40 THEN LET EL=40
```
— Works out last step to list to (EL).
```
260 FOR I=S TO EL
270 PRINT "  ";I;".   ";V$(M(I));
```
— Prints step numbers and commands.
```
280 IF D(I)>0 THEN PRINT TAB(15);D(I);
```
— Prints duration (D) if any. Array D holds the duration for the step.
```
290 PRINT
300 NEXT I
310 PRINT "PRESS RETURN FOR MENU"
320 INPUT Q$:RETURN
```

Enter steps
```
330 CLS
340 PRINT:PRINT "STEP NUMBER":INPUT N
```
— Asks for step number (N).
```
350 IF N=0 OR N>40 THEN RETURN
360 PRINT "COMMAND":INPUT C$
```
— Asks for command (C$).
```
370 LET V=0
380 FOR I=1 TO 17
390 IF C$=V$(I) THEN LET V=I
400 NEXT I
```
— V = Number of position of the command in array V$. Sets V to 0 then compares command (C$) to commands stored in V$. Sets V if match is found.
```
410 IF V=0 THEN PRINT "COMMAND NOT RECOGNIZED":GOTO 360
```
— Command not recognized.
```
420 LET M(N)=V
```
— Records command in array M.
```
430 IF V=10 OR V=11 OR V=16 OR V=17 THEN LET D(N)=0:GOTO 470
```
— Jumps to line 470 for commands which have no repeat or duration.
```
440 IF V<10 OR V=15 THEN PRINT "DURATION":GOTO 460
```
— Asks for duration for relevant commands.
```
450 PRINT "TO WHICH STEP"
```
— Asks where to branch to.
```
460 INPUT D(N)
```
— D(N) = duration for this step.
```
470 GOTO 340
```

Move wheels
```
480 LET G=P(Q):LET N=D(PC):LET TF=0
490 GOSUB 770:RETURN
```
— Array P holds the binary numbers to send to the output lines in the port, for each of the movement commands. G is the number actually sent to the port. TF = 0 means no sensor test is needed as wheels have no sensors.

Arm up
```
500 IF A$="UP" THEN RETURN
```
— If A$ is up, arm won't move.
```
510 GOSUB 580
520 IF SF=1 THEN LET A$="UP"
```
— SF = sensor flag. If arm sensor is on, this means arm is fully up.

*Because it takes a while for the robot to begin moving after a command, the computer waits for a time after the sensors go off before testing to see if they are on again. Both the gripper and arm sensors work in the same way.

```
530 RETURN
```

Arm down

```
540 IF A$="DO" THEN RETURN
550 GOSUB 580
560 IF SF=1 THEN LET A$="DO"
570 RETURN
```

540–570: Same as lines 500-530.

Move arm

```
580 LET TN=128
590 GOSUB 880:LET N=D(PC):LET TF=1
600 GOSUB 770:LET A$=""
610 RETURN
```

- 580: TN tells computer which input line to test.
- 590: Waits for the sensor to go off then checks to make sure it has not come on again before moving arm.
- 600: Moves arm and sets A$ to indicate mid-way position.

Robotrol commands

Open gripper

```
620 IF G$="OP" THEN RETURN
630 GOSUB 660:LET G$="OP":RETURN
```

- 620: Returns if gripper is open.
- 630: Moves gripper and sets G$ to indicate gripper is open.

Close gripper

```
640 IF G$="CL" THEN RETURN
650 GOSUB 660:LET G$="CL":RETURN
```

- 640–650: Same as 620-630

Move gripper

```
660 LET TN=64:GOSUB 880:LET N=1E5
670 LET TF=1:GOSUB 770:RETURN
```

- 660: Waits for sensor to go off then sets duration to large number so gripper moves a long way. TN tells the computer which input line to test.
- 670: TF = 1 therefore sensors must be tested. Moves gripper.

GOTO

```
680 LET PC=D(PC)-1:RETURN
```

- 680: Sets step counter to step before one asked for. When 1 is added (line 210) correct step is obtained.

IFOFF

```
690 LET TN=32:GOSUB 850:IF SF=0 THEN GOSUB 680
700 RETURN
```

- 690: Tests light sensor input line. Calls GOTO routine if sensor flag is off.

IFON

```
710 LET TN=32:GOSUB 850:IF SF=1 THEN GOSUB 680
720 RETURN
```

- 710: Same as line 690 but calls GOTO routine if sensor flag is on.

REPEAT/ END

```
730 LET R=PC:LET E=D(PC)
740 LET C=0:RETURN
750 LET C=C+1:IF C<E THEN LET PC=R
760 RETURN
```

- 730–740: R is a record of step number of command. E = number of repeats wanted. C = count.
- 750: Adds to count. If not the final repeat (E) then sets step number back to repeat statement (R).

Run motors

```
770 GOSUB 840:LET T=0
780 IF TF=0 THEN GOTO 800
790 GOSUB 850:IF SF=1 THEN GOSUB 830:RETURN
800 IF INKEY$(0)<>"" THEN GOSUB 830:PF=1:RETURN
810 LET T=T+1:IF T<N THEN GOTO 780
820 RETURN
```

- 770: Turns on motors.
- 780: Misses out sensor test if not needed.
- 790: Does sensor test. Turns off motors and returns for next step if sensor (SF) is on.
- 800: Turns off motors if a key is pressed, sets panic flag (PF) and returns.
- 810: T counts up to duration (N) specified.

Turn off motors

```
830 ?&FE60=0:RETURN
```

- 830: Turns all motors off.

Turn on selected motors

```
840 ?&FE60=G:RETURN
```

- 840: Turns selected motors (G) on.

Test a sensor

```
850 LET SF=0
860 IF (?DP AND TN)=0 THEN LET SF=1
870 RETURN
```

- 850: Sets sensor flag to 0.
- 860: Sets sensor flag (SF) to 1 if low voltage on sensor input line.

Wait for a sensor to go off

```
880 LET G=P(Q):GOSUB 840
890 GOSUB 850:IF SF=1 THEN GOTO 890
900 FOR T=1 TO 500:NEXT
910 RETURN
```

- 880: Selects motors and turns them on.
- 890: Tests again if sensor still on.*
- 900: Pauses before returning to start move (sensor may come back on momentarily which will stop motors).

Read data

```
920 DIM M(40):DIM D(40):DIM P(11):DIM V$(17)
930 FOR I=1 TO 17:READ V$(I):NEXT I
940 FOR I=1 TO 11:READ P(I):NEXT I
950 RETURN
```

- 920: M stores number indicating command chosen for each step.
- 940: D = duration for each step, P = numbers to send to port for each movement command. E.g. for command B (Backwards) number 7 is sent to port. V$ holds the known commands.

Robotrol commands and numbers to send to port

```
960 DATA "F","B","FR","FL","BR","BL","STOP","UP","DOWN","OPEN","CLOSE"
970 DATA "GOTO","IFON","IFOFF","REPEAT","END","HALT"
980 DATA 6,7,4,2,5,3,0,16,17,9,8
```

"Start-up" procedure

```
990 CLS:PRINT:PRINT "IS GRIPPER OPEN FULLY (Y/N)":INPUT I$
1000 IF I$="N" THEN LET Q=10:GOSUB 630
1010 PRINT:PRINT "IS ARM FULLY DOWN (Y/N)":INPUT I$
1020 IF I$="N" THEN LET Q=9:LET PC=1:LET D(1)=1E5:GOSUB 550
1030 LET G$="OP":LET A$="DO"
1040 RETURN
```

- 1000: Opens the gripper fully if not already open.
- 1020: Moves arm down fully if not already down.

Instructing your robot

The main program is written in BASIC and enables you to use a set of special instructions, called Robotrol, to instruct the robot. You can give the robot up to forty separate instructions. This page shows you how to use Robotrol, with some examples opposite. This is what to do after typing in the main program:

1 Run the program by typing or pressing RUN.

```
IS GRIPPER FULLY OPEN (Y/N)?
IS ARM FULLY DOWN (Y/N)?
```

2 The computer carries out an automatic procedure to find out if the robot is in the correct starting position (i.e. the gripper is fully open and the arm fully down). The questions above appear on the screen. If you answer N (for no), the robot will move to the starting position. Type Y if you have a robot vehicle.

```
1. ENTER STEPS
2. DISPLAY
3. GO
TYPE A NUMBER ?
```

3 The menu above appears on the screen. To give the robot instructions, type 1, and you will see this: `STEP NUMBER ?`

Each instruction you give the robot has a step number. Start by typing 1. You can have up to 40 steps in a sequence of instructions. Change instructions at any time by typing the step number. Next the computer asks you this:

```
COMMAND ?
```

Type one of the commands shown in the list on the right, depending on what you want the robot to do. Next you will see this:

```
DURATION or TO WHICH STEP ?
```

If you type a command to make the robot move or stop, DURATION? appears. Type a number after this question. Low numbers make the robot move a short distance or pause for a short time, high numbers make it go further. Experiment with different numbers for your robot, as distance or time depends on motor speed, gear ratio and type of computer. TO WHICH STEP? appears for other kinds of commands. Type a number after this question, depending on the command.

Next type 0 to get back to the menu. Type 2 to see your instructions, or 3 to make the robot carry them out.

Robotrol commands	
Movement commands	
F	Goes forwards
B	Goes backwards
FR	Goes forwards right
BR	Goes backwards right
FL	Goes forwards left
BL	Goes backwards left
UP	Arm goes up
DOWN	Arm goes down
OPEN	Gripper opens
CLOSE	Gripper closes
Other commands	
GOTO	Goes to a chosen step number in Robotrol instructions.
IFON	If the light sensor is on, goes to a chosen step number in Robotrol instructions.
IFOFF	If the light sensor is off, goes to a chosen step number in Robotrol instructions.
REPEAT	Repeats all commands up to command END, a chosen number of times.
END	Marks the end of a set of commands to be repeated.
STOP	Stops everything for a chosen duration. Use this to get the robot to pause between commands.
HALT	Turns motors off at the end of a sequence of instructions and returns you to the menu.

Arm robot users: the sensors make the motors switch off automatically if either the arm or gripper have moved as far as they will go.

PANIC BUTTON: Press any key to stop the robot if something goes wrong.

Shopping list

Here is a complete list of things you need to build the robot and circuit. There are also details of plugs you need to connect to your computer, interfaces for the Spectrum, and useful addresses of suppliers.

Robotrol examples

Here are some examples of Robotrol instructions. The first set of instructions make the robot follow a thick black line drawn on paper, using the light sensor.

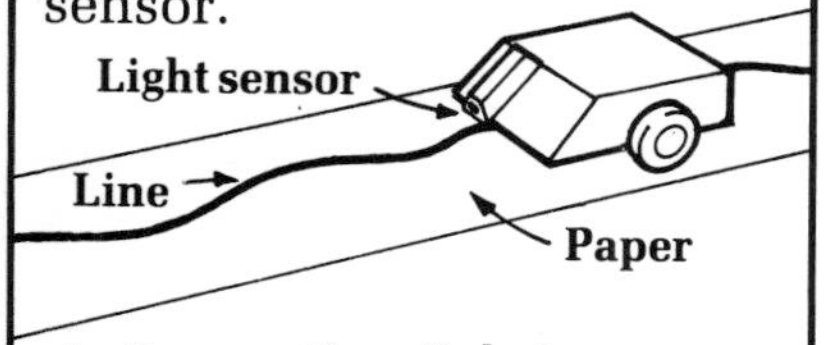

1.	F	10	**Robot goes forwards until it goes off the line.**
2.	IFOFF	1	
3.	REPEAT	10	**Turns right. If it finds the line it goes forwards again.**
4.	FR	10	
5.	IFOFF	1	
6.	END		
7.	REPEAT	30	**Turns left if it can't find the line. Moves forwards if it does.**
8.	FL	10	
9.	IFOFF	1	
10.	END		
11.	STOP	0	**Stops if it can't find the line.**
12.	HALT		

These instructions make the robot go to the end of a straight line drawn on paper, pick up an object, and return.

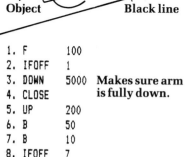

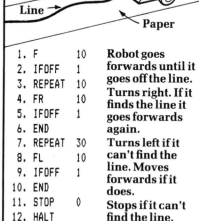

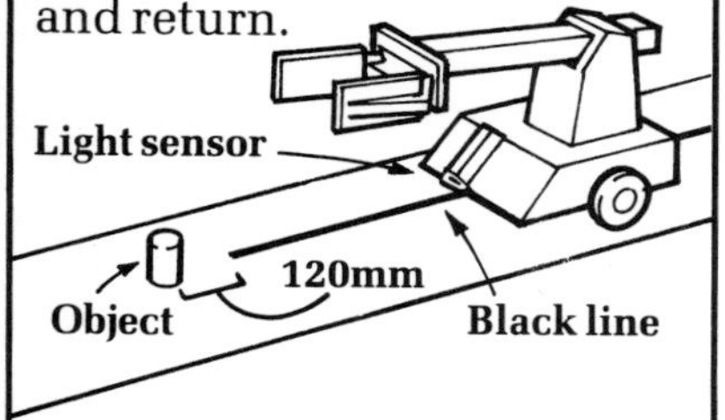

1.	F	100	
2.	IFOFF	1	
3.	DOWN	5000	**Makes sure arm is fully down.**
4.	CLOSE		
5.	UP	200	
6.	B	50	
7.	B	10	
8.	IFOFF	7	
9.	DOWN	500	**Makes sure arm is fully down.**
10.	OPEN		
11.	STOP	1	
12.	HALT		

Tools

If you do not already have them, you can buy the tools in a hardware shop, toolstore, or some timber yards.

Small screwdriver
Metal ruler
Craft knife and blades
Pencil
Paintbrush (15mm decorator's brush will do).
Junior hacksaw (or larger hacksaw, but it is more difficult to use to cut tube).
Tenon saw (or panel saw)
Hand drill (or electric drill)
Pliers (if you use combination pliers you don't need wire cutters).
Wire cutters
Wire strippers
Small file
Scissors
Small vice (or home-made clamp – see page 8).
G-clamp (or home-made clamp).
Soldering iron with small bit
2mm twist drill bit
6mm twist drill bit
Twist drill bit the same size as the outside diameter of tube used to make bearings, or a home-made tube drill (see page 8).

Modeller's materials

You should be able to buy all these things from a model making shop. If not, there are some mail-order addresses at the end of this list.

2 tubes of clear glue or balsa cement
6mm square balsa spar × 3.5m long.
6mm thick balsa sheet 90mm wide × 0.5m long.
3mm thick balsa sheet 75mm wide × 1m long.
1½mm thick balsa sheet 90mm wide × 1.5m long.
6mm diameter (maximum) metal or plastic tube × 1m long.
Metal rod (to fit snugly inside the tube you buy) × 1m long.
Brass shim about 15mm wide × 40mm long. You could use tin foil instead.
10 small self-tapping or wood screws (8 to fit gearbox or motor fixing holes and 2 to screw a connector block to the robot base).
Enamel undercoat
Enamel gloss paint
White spirit or turps substitute
4 small battery-powered motors, between 1½V and 6V power. You only need 2 if making a robot vehicle or stationary arm robot.
4 gearboxes or geartrains to fit your motors and which reduce the speed to between 30 and 150 revolutions per minute.
2 wheels 60mm-100mm diameter.
Fine-grade sandpaper.

Como Drills (for motors and gearboxes), The Mill, Mill Lane, Worth, Deal, Kent, CT14 0PA, England.

W Hobby Ltd (for modeller's materials), Knight's Hill Square, London, SE27 0HH, England.

Shopping list continued over page

Wood

You can buy all the wood from a timber yard or hardware shop. Ask if they have any offcuts as these are cheaper than buying whole pieces.

Robot base: 6-12mm thick plywood or chipboard 200mm wide × 250mm long.
Home-made clamp: 15mm thick softwood, plywood or chipboard 200mm wide × 350mm long.
Workboard: any thickness plywood, chipboard or hardboard about 500mm wide × 750mm long.
Sanding block: any small scrap of wood.

Odds and ends

You may find some of these things around the house. If not, you can buy most of them in a hardware shop.

Sponge (for soldering – any kind will do)
Tin foil about 75mm square
Thin foam rubber about 75mm wide × 300mm long.
Fishing line or strong twine about 1½m long
1 drawing pin
6 metal paperclips
1 steel nail about 75mm long
2 elastic bands about 80mm long when unstretched.
2 compression springs about 30mm long.
2 tension springs, or short strong elastic bands, about 25mm long when stretched.
1 matchstick
1 matchbox
Roll of sticky tape
Roll of insulating tape
Velcro about 150mm long. Velcro is a trade name. It consists of two pieces of stiff fabric, one with tiny hooks, the other with loops, which stick together when pressed.

Electronic components

You can buy the electronic components from a components shop, or by mail-order. An address is given at the end of this list.
Ask in your local TV repair shop or look in the telephone directory to find where the nearest components shop is.
Ask your supplier for substitute components of an equivalent type if they don't have the ones specified here.

Relays: 5 sub-miniature single pole changeover relays, coil voltage 12V d.c. (See very important notes on relays opposite.)
Transistors: 6 × BC108 or BC107.
Diodes: 5 × any diode in the series 1N4001-1N4007.
Resistors: Ask for ¼-½ watt with 5%-10% tolerance. 2 × 330Ω (for LEDs), 3 × 1KΩ, 7 × 2.2KΩ (sometimes written 2K2Ω), 5 × 6.8KΩ (sometimes written 6K8Ω).
Potentiometer: 1 × 100KΩ, LIN or LOG.
Light dependent resistor: 1 × ORP 12.
18-20 SWG cored solder
0.1 inch size Veroboard 31 tracks wide × 60 holes long. (This is enough for an extra piece 10 holes long to practise your soldering.)
10-way ribbon cable × 4m (about 7 × 0.2mm strands in each wire), or "bell wire" × 40m (about 7 × 0.2mm thick strands). Do not use mains cable.
2 connector (or terminal) blocks with 10 pairs of holes.
Batteries: 1 × 12V, 1 × 6V and 2 of a suitable voltage for your motors. See page 32 for notes on using transformers and combinations of lower powered batteries.
DO NOT USE CAR OR MOTORCYCLE BATTERIES OR MAINS ELECTRICITY.

Maplin Electronic Supplies Ltd., P.O. Box 3, Rayleigh, Essex, SS6 8LR, England.

Computer connectors

These are available either from components shops or computer dealers.

BBC: 20-way I.D.C. (short for insulation displacement connector) connector and cable.
C 64/VIC 20: 0.156 inch pitch female edge connector with 24 pins (two rows of twelve).

Spectrum interface

You need to buy a parallel input/output interface board for the Spectrum. Look for advertisements in computer magazines or write to one of the companies listed below.

If you already have a Spectrum sound board you may be able to use it as an I/O interface. Check the manual to find out.

Read the instructions with the interface you buy very carefully. Depending on the type you buy, you will have to use either IN and OUT commands, or PEEK and POKE commands, as shown in the Spectrum program changes on page 35. Check the interface instructions to see which to use. Most interfaces have 8 input and 8 output lines. Use those numbered 0-4 as output, and those from 5-7 as input.

Glanmire Electronics, Meenane, Watergrasshill, Co. Cork, Eire.
William Stuart Systems Ltd, Quarley Down, Cholderton, Salisbury, Wiltshire, SPQ 0DZ, England.

Relay notes

It is very important to use the correct type of relays for the circuit on page 30. There are several makers of the same type and they number the pins differently. Number your pins as shown on the right as these are used in the circuit instructions. Listed below are manufacturers' type numbers and some useful addresses if your components supplier does not have the correct type.

Fujitsu FBR211 series type B or E

RS Components number 348-510

Fujitsu Component Europe, B.V, Rijnkade 19B, 1382 GS Weesp, The Netherlands.

Fujitsu America Inc.
918 Sherwood Drive, Lake Buff, Illinois 60044, USA.

Fujitsu Limited, 6-1, Marunouchi 2-chome, Chiyoda-ku, Tokyo 100, Japan.

Tempatron Ltd., 6 Portman Road, Battle Farm Estate, Reading RG3 1JQ England.

Relay pins

1 Place the pins of your relay over this guide. It will not fit the circuit unless they line up with the dots.

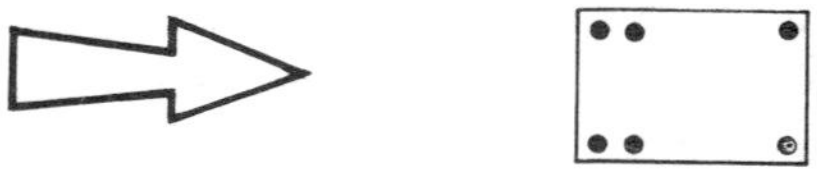

2 Turn the relay on its back. Use the numbers shown to identify the pins.

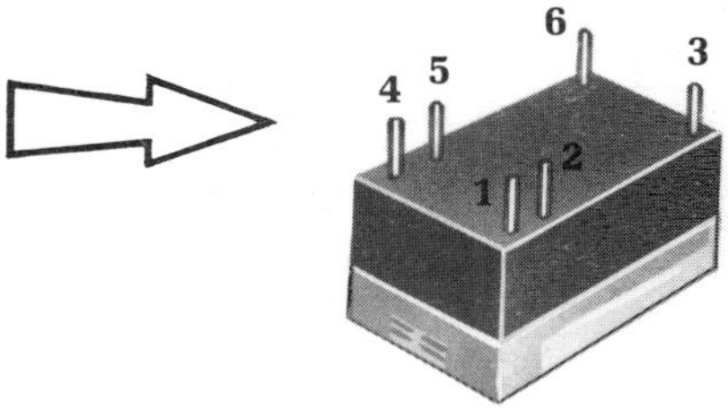

3 You may get a circuit diagram like this to identify the relay pins. Use the numbers shown here. This diagram is a "pin view", which means you identify the pins with them facing you.

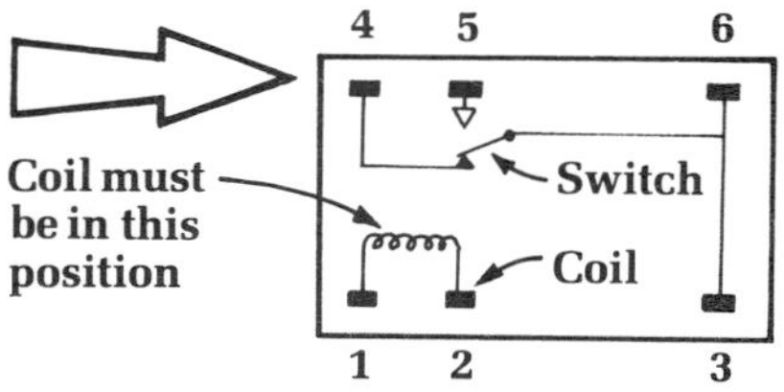

Substitute relays

This is what to do if you cannot get relays with the pins in the correct positions: Look carefully at the relay circuit diagram and substitute the pin numbers with those used above. Then solder short lengths of wire to the pins. Solder the wires into the Veroboard instead of the pins.

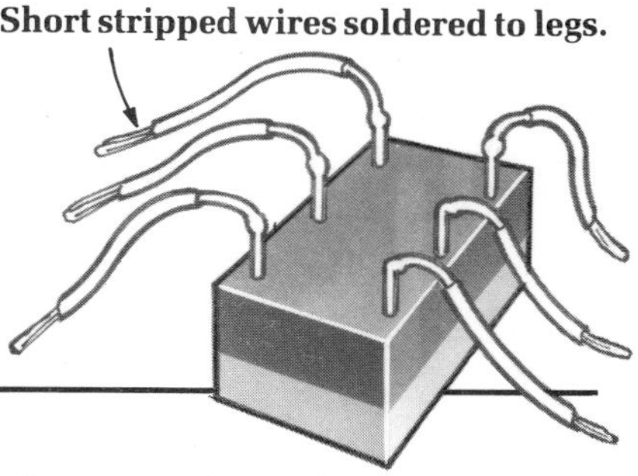

Switching circuit template

Use this template for the circuit on page 30. The rectangles identify where relays go, spots identify where component legs or pins and wires go, and squares show where track breaks are.

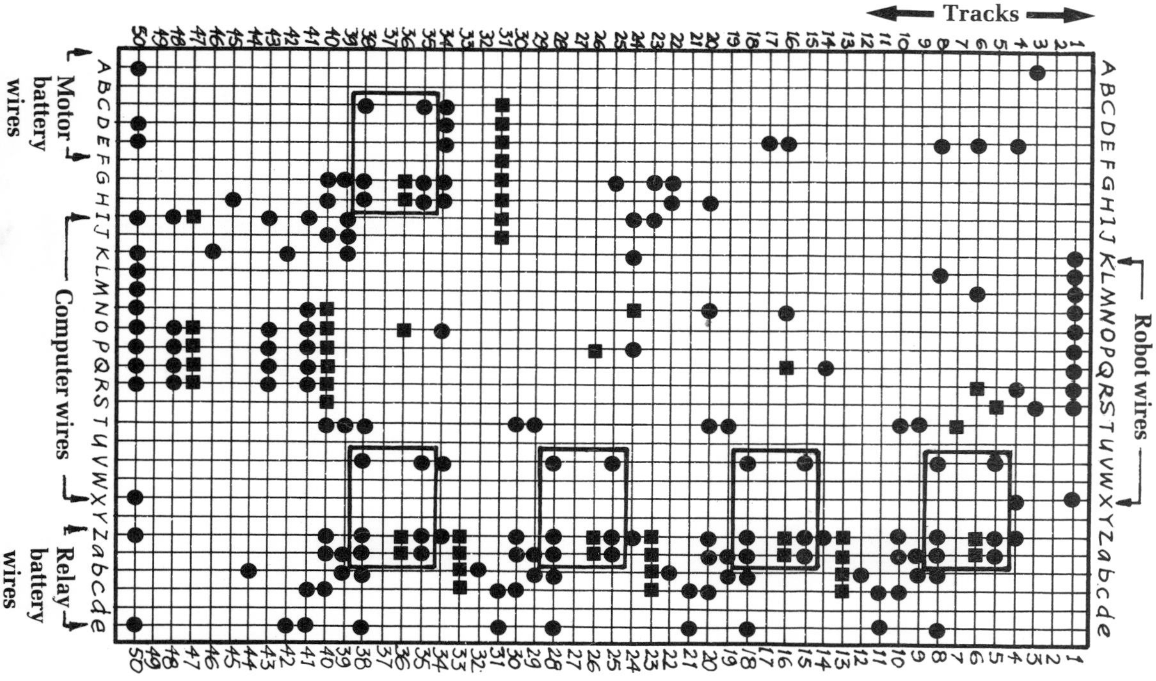

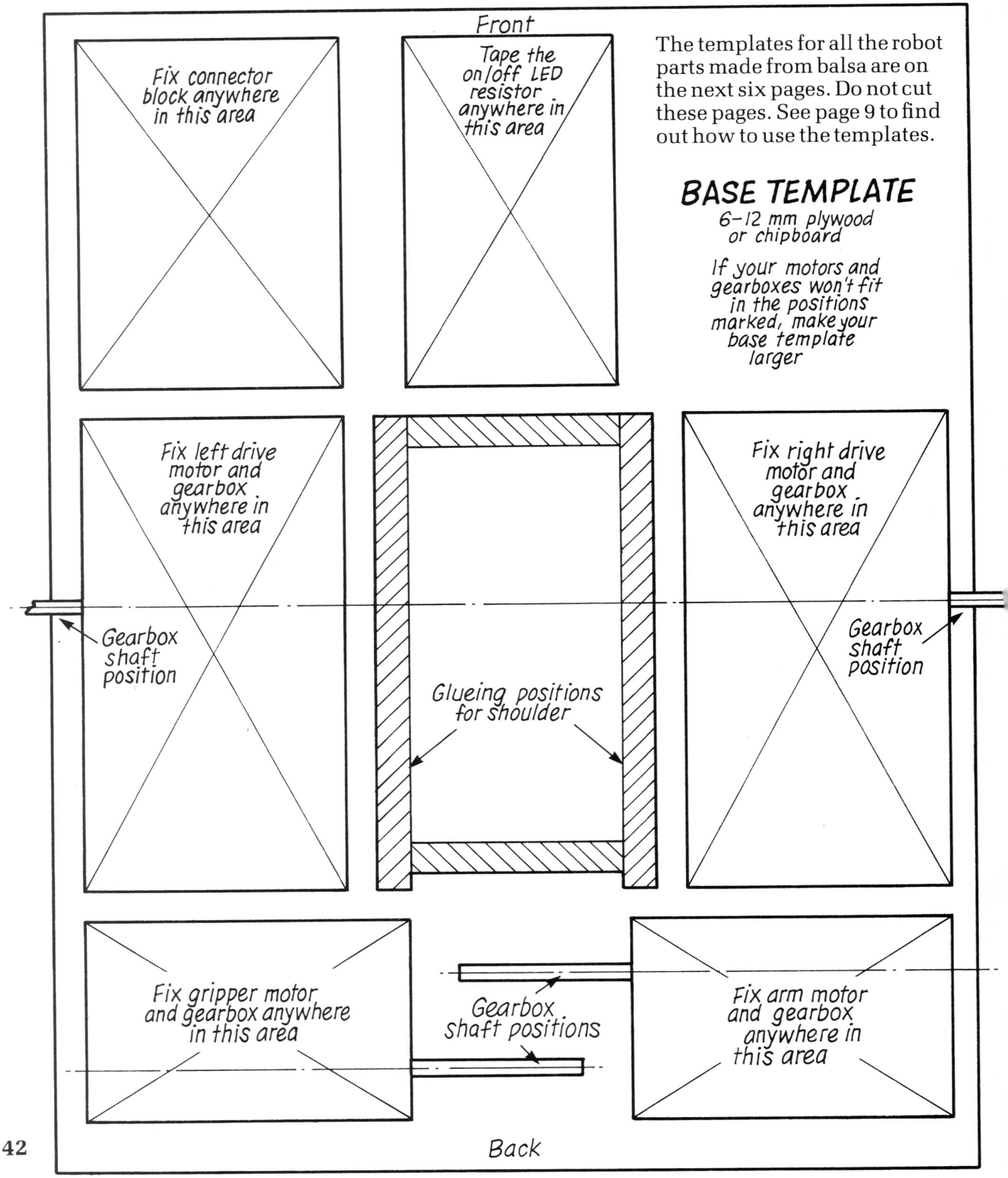
Front
Fix connector block anywhere in this area
Tape the on/off LED resistor anywhere in this area
The templates for all the robot parts made from balsa are on the next six pages. Do not cut these pages. See page 9 to find out how to use the templates.
BASE TEMPLATE
6–12 mm plywood or chipboard
If your motors and gearboxes won't fit in the positions marked, make your base template larger
Fix left drive motor and gearbox anywhere in this area
Fix right drive motor and gearbox anywhere in this area
Gearbox shaft position
Gearbox shaft position
Glueing positions for shoulder
Fix gripper motor and gearbox anywhere in this area
Gearbox shaft positions
Fix arm motor and gearbox anywhere in this area
Back

Shoulder and cover templates

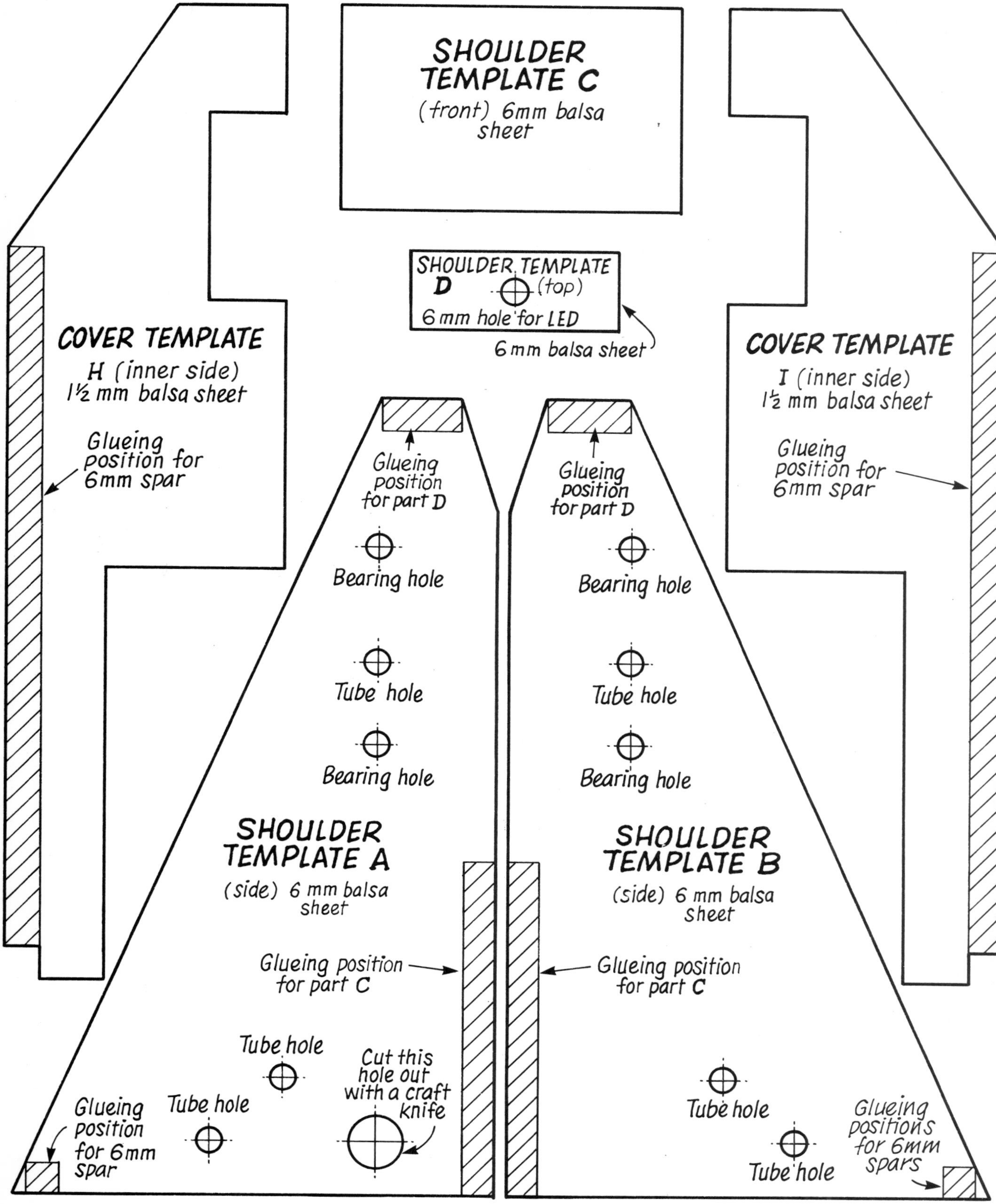

Arm and light sensor templates

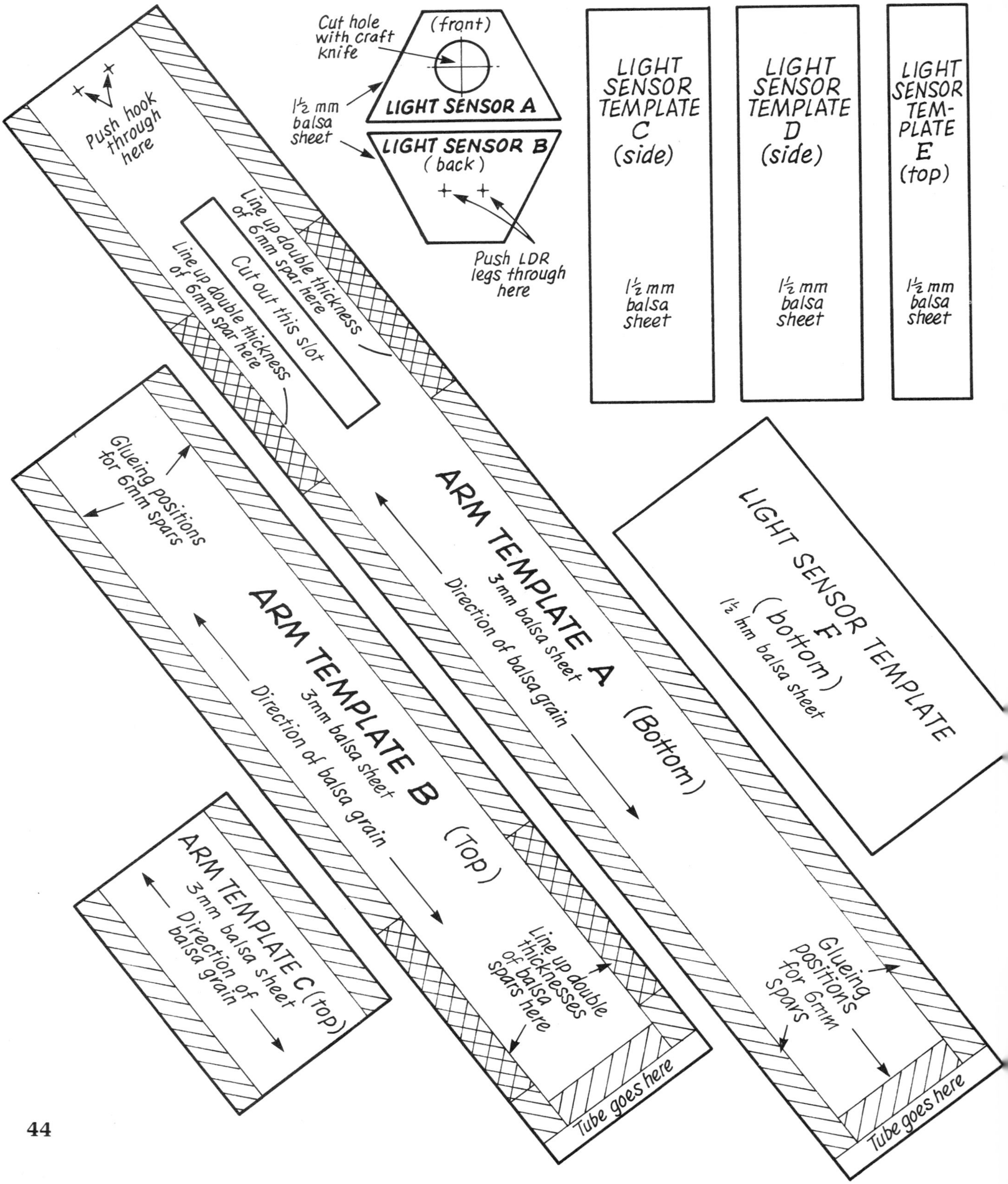

Cut hole with craft knife

(front)

LIGHT SENSOR A

1½ mm balsa sheet

LIGHT SENSOR B
(back)

Push LDR legs through here

LIGHT SENSOR TEMPLATE E (top)

1½ mm balsa sheet

Push hook through here

Line up double thickness of 6mm spar here

Cut out this slot

Line up double thickness of 6mm spar here

ARM TEMPLATE A
3mm balsa sheet
Direction of balsa grain
(Bottom)

Glueing positions for 6mm spars

ARM TEMPLATE B
3mm balsa sheet
Direction of balsa grain
(Top)

Line up double thicknesses of balsa spars here

Tube goes here

ARM TEMPLATE C (top)
3mm balsa sheet
Direction of balsa grain

LIGHT SENSOR TEMPLATE F
(bottom)
1½ mm balsa sheet

Glueing positions for 6mm Spars

Tube goes here

Gripper templates

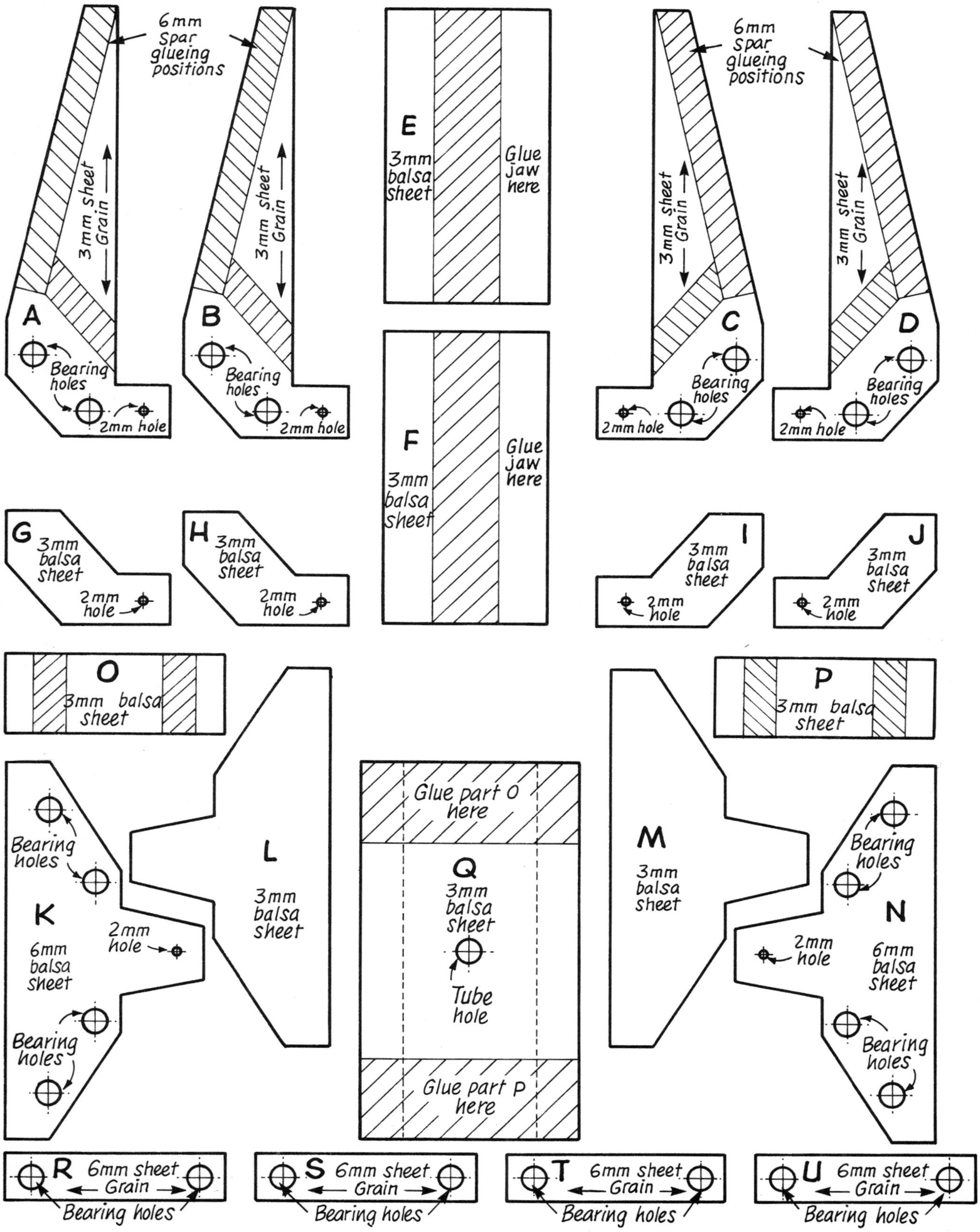

Cover templates

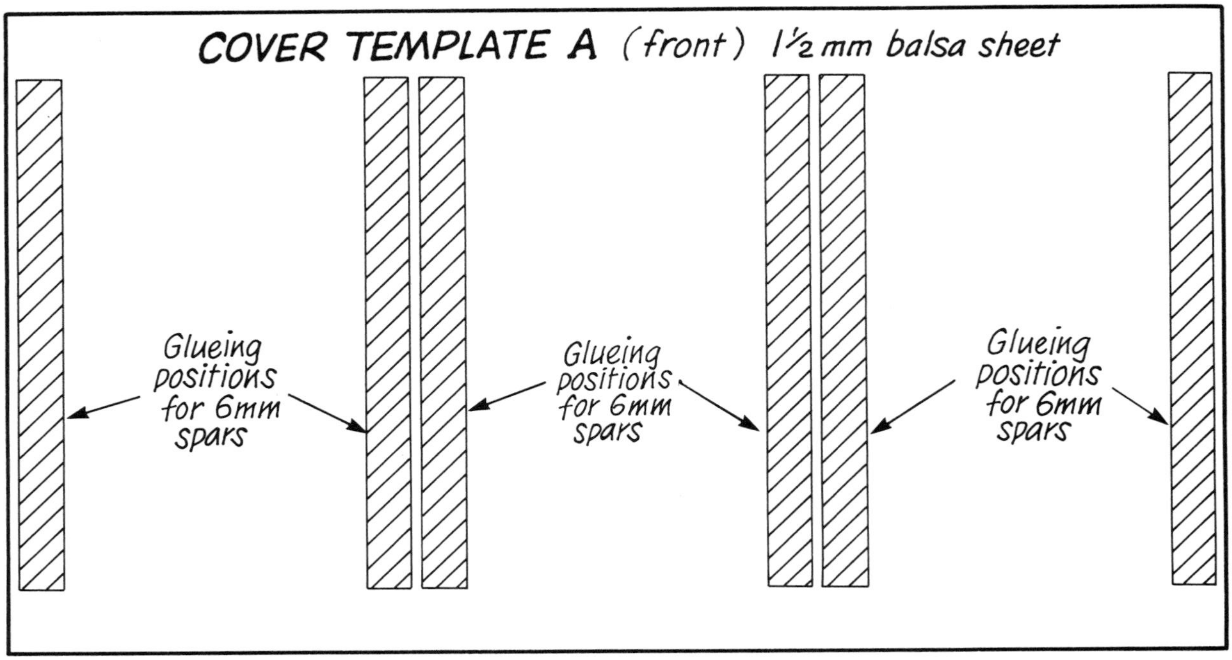

COVER TEMPLATE A (front) 1½ mm balsa sheet
Glueing positions for 6mm spars
Glueing positions for 6mm spars
Glueing positions for 6mm spars

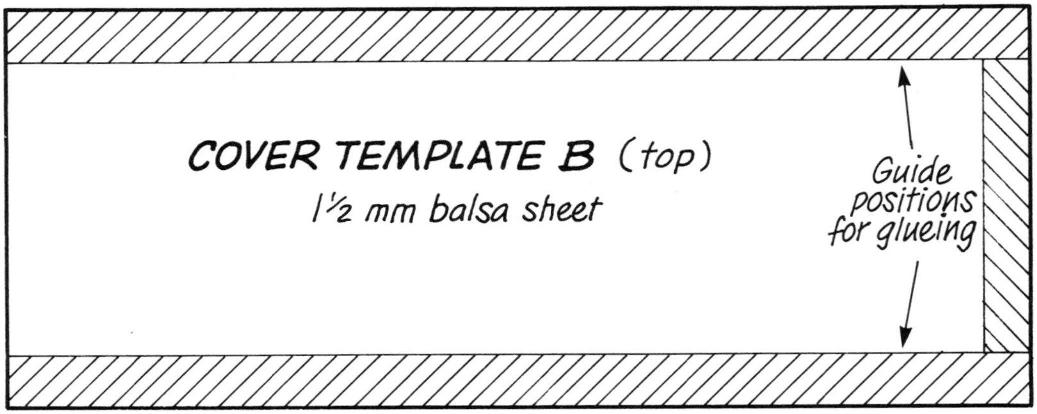

COVER TEMPLATE B (top)
1½ mm balsa sheet
Guide positions for glueing

COVER TEMPLATE C (top)
1½ mm balsa sheet
Guide positions for glueing

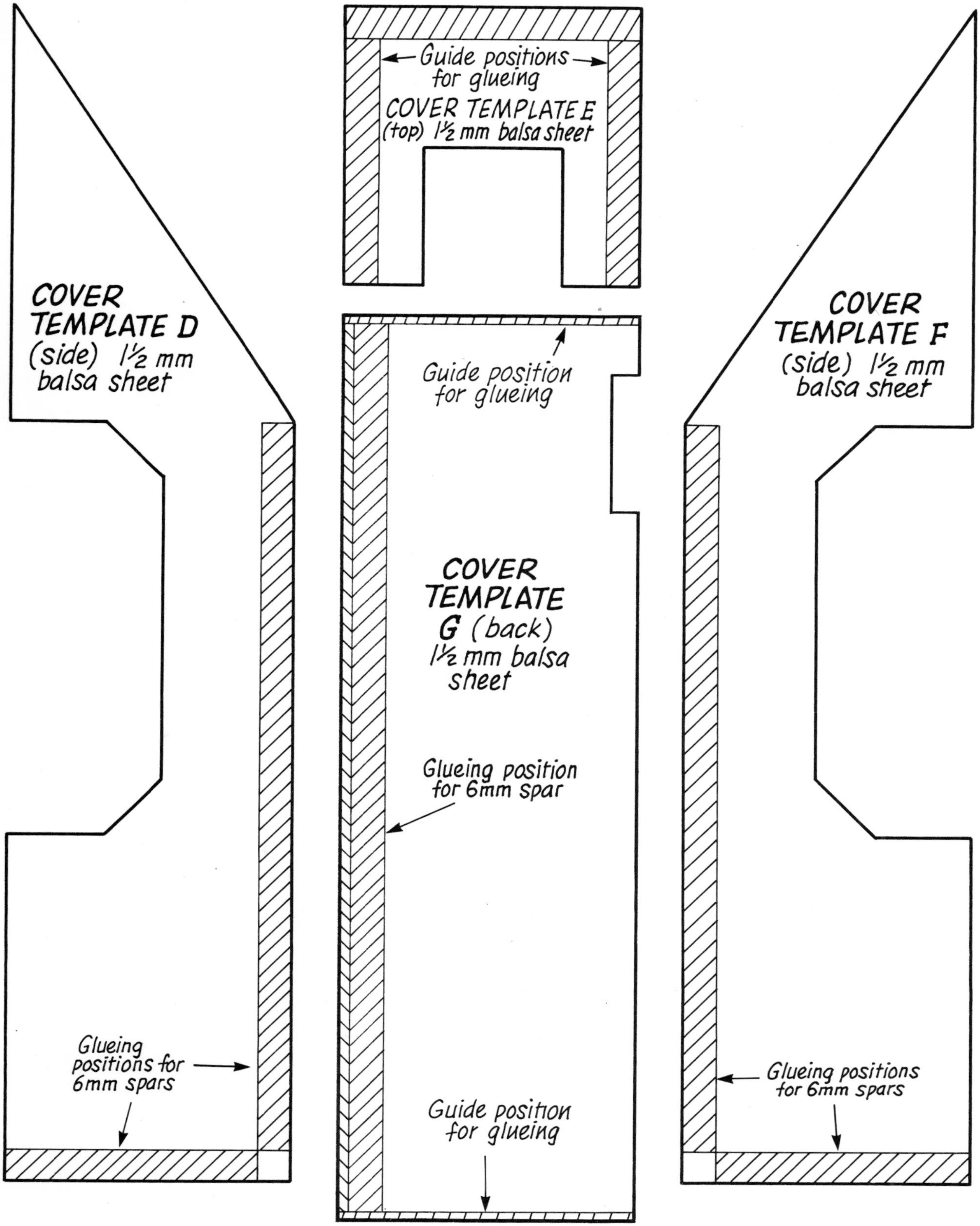

Guide positions
for glueing
COVER TEMPLATE E
(top) 1½ mm balsa sheet
COVER
TEMPLATE D
(side) 1½ mm
balsa sheet
Guide position
for glueing
COVER
TEMPLATE F
(side) 1½ mm
balsa sheet
COVER
TEMPLATE
G (back)
1½ mm balsa
sheet
Glueing position
for 6mm spar
Glueing
positions for
6mm spars
Guide position
for glueing
Glueing positions
for 6mm spars

Index and circuit diagrams

First published in 1984 by Usborne Publishing Ltd, 20 Garrick Street, London WC2E 9BJ, England.

Motor circuit

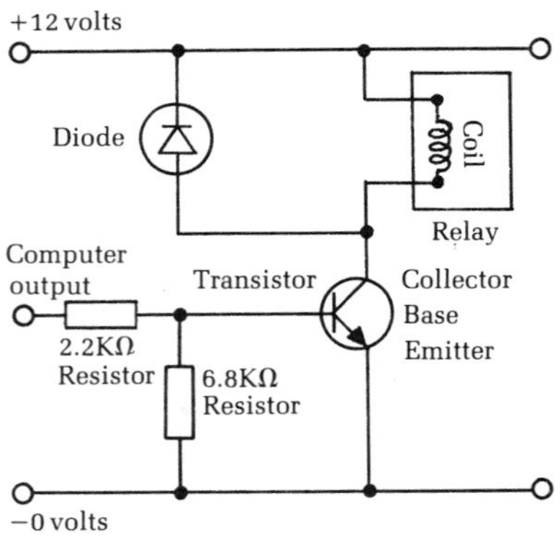

This is the motor control circuit. It is repeated five times in the switching circuit – four times to switch each motor on and off, and once to control their polarity to make them go either backwards or forwards.

Light sensor circuit

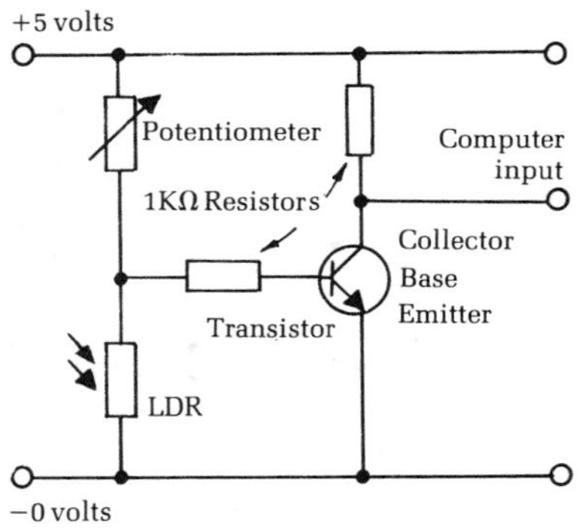

This is the light sensor circuit, and is included with the switching circuit on the same piece of Veroboard.

Test failure

Ask someone else to check the circuit and connections if any test fails, as it is easy to miss something. Use fresh batteries. If the circuit still does not work, pack it carefully with enough stamps for return postage and send it to:
Electronics Advisor,
Usborne Publishing,
20 Garrick Street,
London WC2 9BJ

Answers to questions on page 25: 1 = 2.2KΩ, 2 = 6.8KΩ, 3 = 1KΩ